Design Education Across Disciplines

Miikka J. Lehtonen · Tomi Kauppinen ·
Laura Sivula
Editors

Design Education Across Disciplines

Transformative Learning Experiences for the 21st Century

palgrave
macmillan

Editors
Miikka J. Lehtonen
College of Business
Rikkyo University
Tokyo, Japan

Tomi Kauppinen
Aalto Online Learning
Aalto University
Espoo, Finland

Laura Sivula
Aalto University
Espoo, Finland

Aalto University Executive Education Ltd.
Helsinki, Finland

ISBN 978-3-031-23151-3 ISBN 978-3-031-23152-0 (eBook)
https://doi.org/10.1007/978-3-031-23152-0

This Palgrave Macmillan imprint is published by the registered company Springer Nature Switzerland AG
The registered company address is: Gewerbestrasse 11, 6330 Cham, Switzerland

Foreword

Designing across disciplines is the way to go. Building a sustainable future requires a combination of multidisciplinary collaboration, challenge-based learning, and design thinking. The big societal problems of today's and tomorrow's world cannot be solved in disciplinary silos—instead, we need to work together and combine diverse expertise to create solutions that make change happen. In addition, we need to create those solutions at all levels, from the level of individual products and services all the way to large-scale systemic changes such as circular economy or new energy systems.

However, this is easier said than done. It is also why we urgently need this book. *Design Education Across Disciplines* provides us with the latest ideas and learning on how we can create the capabilities of working across disciplinary boundaries for a better future. It all starts with education by drawing on three perspectives: multidisciplinary collaboration, challenge-based learning, and design thinking. In my dual role as the provost of Aalto University and a management scholar interested in the future of work, I have had the privilege to see in action how the three aforementioned perspectives have been combined in education.

Aalto University was founded in 2010 as a societally embedded and innovative research university for a better world. From the start, its core idea has been to create excellence within and across three fields—technology, business, and art and design. Through education, we want to inspire game changers who are open-minded and take initiative and responsibility, with each making a difference in their own way. Multidisciplinary programmes, courses, and projects have been a key part of this endeavour. Crossing disciplinary boundaries has not been easy, but it has been a hugely valuable learning experience that has created new ideas and ways of thinking that will benefit everyone involved in both tangible and intangible ways. I will share some of my personal learnings below, but before that, a few words on the three key perspectives that I consider foundational in design(ing) across disciplines: multidisciplinary collaboration, challenge-based learning, and design thinking.

First, although I personally like to use the term multidisciplinary collaboration to encompass different ways of working together across disciplinary boundaries, it does in reality involve many different forms and depths. Also terminology varies from inter- to transdisciplinarity and everything in between. I will not take a stand on which terminology should be used, but rather emphasise that different levels of collaboration are needed. For some, creating T-shaped understanding of ways of thinking and working in other disciplines is all that is needed; others want to combine in-depth knowledge of two fields to innovate at their cross-section; and yet others build their core competence around understanding and facilitating cross-disciplinary work. All of these are valuable. I encourage readers to reflect on the different examples of collaboration across disciplines shared in this book, to see "what works for me." For example, in Chapter 4 Keane and Yeow provide insights on how higher education institutions could collaborate with other actors (e.g. public and private sector) in creating research and teaching initiatives with capabilities to respond to contemporary and future challenges.

Second, bringing people from different disciplines together is, however, not sufficient just in itself. The ways of thinking and working are often so different that you very quickly start to experience misunderstandings, friction, and even conflicts. Multidisciplinary collaboration has to be combined with a challenge-based approach, in which we start with a real-life challenge—a goal that we want to achieve or a problem that we want to solve—and work backward from there. Challenge-based learning emphasises active experimentation, where the learners are in charge, and is, therefore, well suited to working across disciplinary boundaries to solve complex problems and create innovative solutions that work. In Chapter 8, Dickson-Deane et al. focus on how learning that takes place in the classroom can be effectively utilised in the workplace. The authors investigate work-integrated learning from multiple perspectives (i.e. higher education, society, industry), which provides excellent reading from this perspective.

Third, and importantly, multidisciplinary collaboration and challenge-based learning need to be combined with design thinking—an iterative process for redefining problems and understanding user behaviours. Traditional linear approaches will not work with the complex systemic problems we are currently facing. Instead, design thinking methodologies will give us tools for challenging assumptions, thinking about problems in new ways, and developing novel solutions. This book is an invaluable resource for understanding how design thinking has been and can be used at different levels. For example, in Chapter 14 Yeo focuses on sustainability from the perspective of the kind of skills future graduates might need in order to thrive in the future workplace. This chapter addresses the future of design education from the perspective of educators and learners, and this chapter can also be implicitly understood as an invitation for other disciplines to collaborate with design educators and practitioners.

My personal experiences with multidisciplinary collaboration, challenge-based learning, and design thinking are from Aalto University, where building these three skills has been an important part of our educational strategy since 2010. We have learned that you have to build these skills at different levels and in many parts of the educational curriculum. We have project courses, in which multidisciplinary teams solve real-life challenges. We encourage all students to take courses from other fields in order to be exposed to, and learn from, how people from other disciplinary backgrounds approach issues. We have built multidisciplinary majors and minors, in which challenge-based learning and design thinking are central learning methods. We are now experimenting with a co-educator team which seeks to integrate sustainability, radical creativity, and entrepreneurial thinking skills into all programmes. We have also incentivised collaborative initiatives with seed funding, built multidisciplinary collaboration into major university research initiatives, and facilitated networking across disciplines around topical themes. We are showcasing our results through transmedia storytelling and in our "Design for Cooler Planet" exhibition which is part of the annual Helsinki Design Week. In the 2021 Design for a Cooler Planet exhibition, for example, we showcased radical ideas that contribute to a resource-wise future. Many of the showcased innovations originated from student projects, which tackled the big challenges of the world and required multidisciplinary collaboration and design thinking. Examples included "leather" made from flower waste, biobased wrapping for cucumbers, and other fascinating biomaterial, sustainable fashion, and environmental living experiments.

The progress we have made is exciting, but it also has not been easy. To me, the crucial first step is to open up mindsets. In the beginning of our journey as a new university, most community members were relatively disinterested in multidisciplinary collaboration or in adopting new challenge—and design thinking—based ways of working. "What's in it for me?" was a frequent implicit question. Opening up mindsets was gradual, and required concrete examples of what these multidisciplinary, challenge-based, and design thinking approaches could look like, and early success stories to show how individual researchers and teachers could benefit from them. Students were much more open-minded, but also their choices were driven by rational thinking. Crossing disciplinary boundaries did not happen automatically for them either. We needed first to make it possible and easy to take the new courses and include them in study plans, and then gradually make them an integrated part of curricula. We are still working on this.

In terms of mindsets, we are now at a stage where most Aalto University's community members believe in the value of multidisciplinary collaboration, and in challenge-based and design thinking approaches but are still learning how to do them in practice. Sharing practical examples such as those presented in this book is invaluable, because examples inspire others to try, give ideas for what to do in very practical terms, and show what you need to think about and what potential pitfalls you need to be mindful about.

Making progress also requires academic champions who are leading the way. The editors of this book—Miikka J. Lehtonen, Tomi Kauppinen, and Laura Sivula—are prime examples of such champions. They have in their own work been living examples of how to work and educate in a way that crosses disciplinary boundaries, starts with a challenge in mind, and utilises design thinking. The fruits of their and other trailblazers' collective learnings can be found within the covers of this book. The authors of the different chapters of this book have years of combined experience in design(ing) across disciplinary boundaries and they have graciously shared both their successes and challenges and the insights they have picked up along the way.

Finally, The World Economic Forum has in their 2020 Future of Jobs Report listed both complex problem-solving and creativity, originality, and initiative among the top skills for 2025, and I agree. The combination of multidisciplinary collaboration, challenge-based learning, and design thinking go a long way in building these future skills. I encourage everyone to read this book with a learning mindset.

Kristiina Mäkelä
Provost
Aalto University
Aalto, Finland

Professor, Department
of Management Studies
Aalto University School
of Business
Aalto, Finland

Preface

The origins of this edited volume go back to 2018 when we, the editors, started discussing the possibility of drafting a book proposal that would take stock of our teaching experiences spanning multiple years and continents. Gass (1972, p. 10) highlights that "creative change in university education and research calls increasingly for an interdisciplinary approach to teaching", and this is exactly one of the reasons for curating this edited volume. With contributions from eleven countries and multiple disciplines, we wanted to bring together diverse voices so as to contribute to the current body of knowledge on transdisciplinary teaching and learning by engaging in a conversation across this volume's chapters.

As such, and inspired by Boling et al.'s (2016) curatorial process, this edited volume consists of three sections that shed light on separate, yet interrelated topics, critical to the future of transdisciplinary teaching and learning in universities. To this end, contributions were selected so that they highlight the plurality of design interventions whilst ensuring theoretical rigour. More specifically, whilst each chapter can be read as a stand-alone entity, we have aimed at increasing the relevance of each chapter by connecting them to other chapters and broader issues through the synthesis sections. We are thrilled to have had prolific educators write these syntheses.

First, each part of the edited volume has been synthesised by engaging in a dialogue with each chapter in its respective part. This dialogue was partly framed based on Shulman's (2005) notions of signature pedagogies. By doing so, each part can be read individually and by focusing on actionable insights, we hope to have ensured each part's theoretical and pedagogical relevance. Second, we also applied Shulman's (2005) framework in the conclusion chapter by offering concrete avenues for further research and strategic pedagogical interventions. Finally, to ensure coherence in the contributions, we also hosted co-design workshops with authors of chapters during the curatorial process so as to enable cross-pollination and dialogue between contributions. In addition, chapter authors also peer-reviewed each other's chapters, and we are truly thankful to everyone for taking the time to bring additional depth and dialogue to the chapters.

Finally, we would like to acknowledge and thank certain people without whom this edited volume would not have been possible. Rebecca "Becky" Wyde supported and encouraged us in the beginning of this process when we were still drafting the proposal. We hope you enjoy, as much as we do, seeing this edited volume published, Becky. In addition, Alice Green's editorial guidance in the latter part of the project was invaluable and Aishwarya Balachandar guided us through the practicalities throughout the project. Thank you so much! Most importantly, we are extremely grateful to all the authors who have contributed to this book. Without you this would not have been possible, so thank you for your contributions as well as for working with us. We have very much enjoyed this process and hope you, too, will enjoy seeing this book published. Albert Treacy did stellar work in proofreading the manuscript—thank you so much for ensuring this book is a delight to read. We would also like to thank Rikkyo University for their financial support in ensuring the final result will be of high quality and Annika Leppäaho for creating a coherent visual identity for this edited volume.

Tokyo, Japan — Miikka J. Lehtonen
Helsinki, Finland — Tomi Kauppinen
Helsinki, Finland — Laura Sivula

Contents

Notes on Contributors

Chantal H. T. A. Brans is the Program Manager Education Innovation of TU/e innovation Space. With a broad background in educational innovation, she is always initiating new projects, research, and best practices. The aim of this effort is to create more impact from innovations.

Claire Brophy is a Lecturer in the School of Design at Queensland University of Technology (QUT), Brisbane, Australia. She has an educational background in industrial design and journalism and researches, writes, and teaches about people's everyday interactions with technologies, products, services, and systems. She explores how sociocultural factors such as an individual's values and experiences influence the design and use of products and systems.

Dean Brough is Discipline Leader in the QUT School of Design and Work Integrated Learning Coordinator for the Faculty of Creative Industries, Education and Social Justice. Dean's applied research centres on the intersection of health and design, with a focus on performance and protective apparel for sub-tropical climates. With over 30 years of tertiary teaching and a wealth of educational leadership and industry expertise, Dean is also a Design Institute of Australia Award recipient.

JiaYing Chew is a Doctoral Researcher at the University of the Arts London (LCC), and a Teaching Assistant at the National University of Singapore. A self-proclaimed multi-disciplinarian, she has a background in project and facilities management that shaped her interests in integrated systems and businesses, while her industrial design training has honed her sensibilities to form and function. Her current research interests focus on the role of design in transdisciplinary higher education.

Camille Dickson-Deane researches the online learning space and how individual differences influence the context, design, functionality, and ability to afford learning. She is a scholar-practitioner in the field of learning sciences and promotes the importance of a 360-degree lens when designing for learning and performance.

Melanie Finger Senior Lecturer in Transdisciplinary Design at the School of Design, QUT, has 20 years of fashion industry experience in local, national, and international clothing design and manufacture. Her teaching focuses on product design and transdisciplinary authentic learning through Impact Lab units. Melanie excels at connecting academia with industry, creating job opportunities, and fostering long-term partnerships. She has led two nationally funded exchanges to South Korea fostering emerging designers.

Threase Finnegan-Kessie holds a Ph.D. in Anthropology from Maynooth University. She completed a Post-Doctoral Research Fellowship with University College Dublin conducting ethnographic research into the User-Centred design of Connected Health technologies. While managing a €2.1 million Horizon 2020 funded European Industrial Doctorate programme, Threase began to research and publish on innovation within the Higher Education sector, with a focus on the growing importance of interdisciplinary in this domain. Threase is a keen traditional flute player and mother of three young boys.

Gorm Gabrielsen is Professor emeritus of Mathematics and Statistics at Copenhagen Business School. His research concerns the development of research designs in general and particular the development of methods for measurements of preferences for single individuals. In addition, he has worked with other specialists and developed statistical methods in business, design, sensory, forensic medicine, criminology, and psychiatry. Gabrielsen has taught statistics in several business studies programmes at Copenhagen Business School and the University of Copenhagen.

Keith Heggart is an early career researcher with a focus on learning and instructional design, educational technology, and citizenship education.

He is currently exploring the way that online learning platforms can assist in the formation of active citizenship amongst Australian youth. Keith is a former high school teacher, having worked as a school leader in Australia and overseas, in government and non-government sectors.

Gloria Hoyos Bustamante holds a Master's degree in Habitat and a Ph.D. in Design and Creation. She has been a professor at the Universidad Nacional de Colombia since 2010 and has served as Academic Vice-Dean. She is currently coordinating the Interdisciplinary Courses of the School of Architecture and Urbanism to strengthen the relationship with the social and business environment. Her interests include teaching and interdisciplinary work related to architecture, territory, habitat, landscape, and bioclimatic architecture.

Tokushu Inamura is a Designer/Engineer, appointed as Assistant Professor of Design Strategy at the Faculty of Design and Center for Designed Futures of Kyushu University. Current research focuses on Post Human Centered Design and Circular Design. He is currently teaching Strategic Design at postgraduate level, and Design Futures at undergraduate level, as well as engaged at the Robert Huang Entrepreneurship Center. He engages in praxis to educate internationally minded transdisciplinary teams to become strategic and holistic innovators. Born in Aotearoa New Zealand, he has lived in the US, Switzerland, and the UK, and he currently resides in Fukuoka City Japan.

Derek Jones is a Senior Lecturer in Design at The Open University (UK) and part of the OU Design Group. His main research interests are the pedagogy of design and creativity, embodied cognition in physical and virtual environments, and theories of design knowledge. Derek is a qualified architect with 15 years of experience in the construction design and procurement industries and is the Convenor of the Design Research Society's Education special interest group.

Tomi Kauppinen is the head of Aalto Online Learning—Online Hybrid Lab at Aalto University. He holds a habilitation (2014) in geoinformatics from the University of Muenster in Germany, a title of docent (2014), a Ph.D. (2010) in media technology from the Aalto University, and M.Sc. (2004) in computer science from the University of Helsinki. His transdisciplinary research investigates cognitive and ubiquitous systems, information visualization, spatial thinking, and online learning. Since 2016, he has led the Aalto University wide strategic initiative, Aalto

Online Learning, which covers activities ranging from blended learning to fully online textbooks and exercises, and from artificial intelligence-based recommendations and assessment to interactive visual simulations.

Kate Keane is a Senior Lecturer and Course Director in Postgraduate and Professional Programmes at the Canterbury Christ Church Business School. She completed an industry-sponsored M.Sc. in Projects and Enterprise Management at UCL following a career in senior management, responsible for investment banking corporate real estate. Kate subsequently led major rail infrastructure projects before embarking on her current role in Higher Education, where she has attained Fellowship of the Higher Education Academy.

Veronika Kelly is an Associate Professor of Design at the University of South Australia. Kelly's research focuses on design culture and design education, with a specific interest in the relationship between communication design and the art of rhetoric. Her work explores design as a social construction, the ways in which design discourse shapes design cultures, and how this in turn informs what design education for contemporary practice is or could be.

Joshua Korenblat is an Assistant Professor of Graphic Design at the State University of New York at New Paltz. Joshua teaches courses about working with complex information to tell visual and verbal stories. Joshua is a Reviews Editor at *Design and Culture*. Earlier in his career, Joshua worked at *National Geographic Magazine* and *Science News Magazine*. Joshua is a cofounder and Art Director-At-Large at Graphicacy, a data visualization firm in Washington, DC.

Tore Kristensen is Professor emeritus in marketing and design at Copenhagen Business School. His research has been concentrated on product development and how the products were perceived. A large part of his work has concentrated on the collaboration between various professions and crafts concerning environments and various artefacts. Several studies concern the value of design, in particular how the economic value guides the perception of products. He has taught various courses in the business of design and stated the integrated design with collaboration between the business school, Denmark's Design School, Danish Technical University, and Academy of Art.

Miikka J. Lehtonen is a multidisciplinary academic and one of the co-founders of the award-winning Nordic Rebels movement. He is currently working at Rikkyo University College of Business as a Specially Appointed Associate Professor. Prior to joining Rikkyo, Miikka has worked at Dubai Institute of Design and Innovation in the UAE, Aalto University in Finland, and the University of Tokyo in Japan. He obtained his Ph.D. from Aalto University School of Business in 2014. His current research interests and teaching methods focus on visual methodologies and knowledge production, management, and creativity in game companies, design practices and management in contemporary organizations, and (blended) design pedagogies.

Deanna Meth is a Senior Lecturer in Curriculum and Learning Design at QUT. She has experience leading and supporting strategic projects in learning, teaching, and research in universities in the UK and South Africa and her recent research has focused on enabling curriculum design, and curriculum development activities, with a focus on transdisciplinarity, authenticity, and student–staff partnership working.

Isabelle M. M. J. Reymen is the Scientific Director of TU/e innovation Space and a full Professor in the field of innovative ecosystems. She combines entrepreneurial thinking with a passion for design processes and aims for system innovation. In 2020 she became Comenius Leadership Fellow and was granted a scholarship for performing research on education innovation with a proposal entitled "Fostering Challenge-Based Learning through TU/e innovation Space". In 2021 she, and her team, were granted the first Higher Education Award in the Netherlands for educational innovations in the past few years.

Aidan Rowe is Associate Professor in Design Studies and Chair of the Department of Art and Design at the University of Alberta, Canada. His research, curatorial and practice interests are in design, health, and education. Recent practice-based work explores future design scenarios with a focus on employing participatory design approaches—working with people—to improve healthcare practices, processes, and services. Recent pedagogic work explores the application of design methods and processes to public health and involves the development of curriculum for medicine and health students.

Laura Sivula is an education thought leader and futurist, education designer, and lecturer, with several active affiliations. She has witnessed the changing landscape in business and education in her role as Business Area Director for both Aalto University and Aalto University Executive Education Ltd., which gives her a unique perspective on challenge-based and life-wide learning. Laura is a sought-after keynote speaker and lecturer on the topics of the future of work, learning organizations, and digital business. Recently, Laura has developed models for life-wide learning across disciplines and platforms, on both national and international levels.

Aldo Valencia is a Lecturer of Design Innovation and in charge of the Entrepreneurship, Design and Innovation Centre at Maynooth University, Ireland. He collaborates with Mi:Lab, a Design Innovation centre, that explores ways to tackle challenges within higher education through the design approach.

Robert Vanderburg (Senior Lecturer—Education, CQUniversity Australia) is an experienced educational psychology and digital pedagogy researcher and has won multiple teaching awards across both America and Australia. His work on developing an innovative digital-based pedagogy for numeracy and literacy has enabled hundreds of pre-service teachers to obtain registration.

Trevor Vaugh is an Assistant Professor of design innovation at Maynooth University and an expert in human-centred innovation, strategy, and design. In 2018, Trevor founded the Maynooth University Innovation Lab (Mi:Lab) with the support of the Higher Education Authority of Ireland. Across his career in innovation, Trevor has accumulated a portfolio of over 50 patents. Recently, Trevor appeared as an expert on the critically acclaimed RTÉ series The Big Life Fix.

Jan D. Vermunt is Professor of Learning Sciences and Educational Innovation at Eindhoven University of Technology, Eindhoven School of Education. His expertise area is the learning sciences, with a focus on teaching and student learning in higher education, and teachers' learning and professional development. Current research interests include effective ways to support personal learning pathways, promoting deep and self-regulated learning in students and teachers, developing innovative teachers, and new pedagogies for challenge-based learning.

Dilys Williams FRSA holds a UAL professorship in Fashion Design for Sustainability and she is the founder and Director of Centre for Sustainable Fashion, University of the Arts London Research Centre, based at the London College of Fashion. Dilys' work explores fashion's relational ecological, social, economic, and cultural elements to contribute to sustainability in and through its artistic, business, and educational practices. She is currently acting as a Special Adviser for All-Party Parliamentary Group (APPG) on Ethics and Sustainability in Fashion and is a member of the jury for Vogue YOOX Challenge—The Future of Responsible Fashion.

Jesvin Puay-Hwa Yeo is Associate Professor in the School of Art, Design, and Media at Nanyang Technological University, Singapore. Her research focuses on how design affects societies and how these societies change the design. Jesvin is the author of award-winning art books: "Vanishing crafts", "Architectural decoration: negotiating symbols across time and place", and "Visual/Senses". Her cultural-related works have been published and exhibited internationally, including at the Basel School of Design, Berlin Communication Museum, and Taiwan Design Museum. These works have also won prestigious design awards, such as the Red Dot Award, GOOD DESIGN™ Award, and iF Communication Design Award.

Pamela Yeow is a Reader in Management and Assistant Dean (External Engagement) at Birkbeck, University of London. Pam completed her Ph.D. at the Institute of Work Psychology, University of Sheffield, on managing effective change strategies in multinational organizations. Recently published articles include a longitudinal study on building sustainable careers for musicians, a case study analysis into how ethical consumerism can be embedded and exploring multiplex networks, leadership, and group performance. She was recently awarded funding grants to understand the differing householder roles in effecting sustainable change.

List of Figures

List of Tables

CHAPTER 1

Introduction

Miikka J. Lehtonen

Since 2018 several issues and events on a global scale have had a profound impact on teaching and learning in higher education institutions. Racial injustice (Dar et al., 2021), Russia's war on Ukraine (Kismödi & Pitchforth, 2022), the Covid-19 pandemic (Jones & Lotz, 2021), regional and global environmental disasters (IPCC, 2022), and deteriorating working conditions in academia (e.g. Hill, 2012) to name but a few examples. While some of these have had more direct impact on teaching and learning than others, Fleming (2021, p. 2) reminds us that the issues go deeper: "[t]he founding mission of public higher education has been pulverised over the last 35 years as universities morphed into business enterprises obsessed with income, growth and outputs" (see also Fleming, 2020). Similarly, as Dar et al. (2021, p. 696) elucidate, "capitalism in all its forms (colonial, slave, mercantile, financial, market, information) has used racist logic to demarcate superior white humanity from deficient Black non-humans, distributing wealth along a racialised scale of human

M. J. Lehtonen (✉)
College of Business, Rikkyo University, Tokyo, Japan
e-mail: miikka@rikkyo.ac.jp

M. J. Lehtonen et al. (eds.), *Design Education Across Disciplines*,
https://doi.org/10.1007/978-3-031-23152-0_1

value" (citing Mills, 1997). While universities have become sites of profit-maximisation where exclusion often seems to be the norm (Harley & Fleming, 2021); the landscape of higher education is becoming more and more diverse. This is evident, for instance, in higher education institutions critically examining their curricula to make them more inclusive and simultaneously equipping students with skills relevant for the future while maintaining academic integrity. In order to understand the new challenges and complexities of society, universities are transforming from information repositories to places where students can learn by solving real-life challenges (e.g. Lehtonen et al., 2022).

Why This Edited Volume? Why Now?

While reading Dewey (1997), Jantsch (1972), or Piaget (1972) one might reach a cynical conclusion that hardly anything has changed during the last decades in terms of how higher education institutions continue to organise teaching in artificial disciplinary silos in the twenty-first century. Although to an extent, educators in higher education institutions continue to be pushed towards delivering courses where participants are counted in hundreds, at the same time there are also examples that echo both experiential learning and transdisciplinarity (see also Gass, 1972). Yet, we believe this edited volume to be timely for two reasons: first, access to higher education is becoming increasingly exclusive (Mitchell, 2022) and those who can afford to educate themselves are often exposed to curricula where key performance indicators matter more than meaningful experiences (Fleming, 2021), and second, the problems we are currently facing and continue to face in the future demand radical transformations in how we engage in teaching and learning in higher education institutions. Having said that, while this volume focuses on universities, we believe the following chapters will also prove useful in other educational institutions and organisations interested in lifelong learning.

Introducing the Different Sections

Part I—Higher Education Leadership and Programme Management Perspective

The higher education institution of the future is forward-thinking and capable of making agile adaptations as the world changes. In this landscape, leadership and managerial capacities are critical in terms of steering and supporting not only top-down but perhaps more importantly, bottom-up transformation. As it is now, however, amidst burdensome administrative duties and increasing attention towards conforming to global rankings and political agendas, disruptive practices are either not encouraged or do not attract enough attention. Due to design's inherent future-creating nature, pedagogical interventions and initiatives drawing on design stand in a pivotal role when it comes to envisioning the future of teaching and learning (e.g. Boling et al., 2016).

We consider that the higher education institution, as a spatio-temporal environment, can enable aspiring academics and managers to develop disruptive practices and use design as a transdisciplinary catalyst. To do so, institutional enablers and barriers need to be understood and acknowledged. Thus, this part of the edited volume sheds light on design-driven pedagogical transformation from the managerial perspective, namely, detailed design cases, discussions on quality and efficiency, and reflections from the management's point of view.

Chapter 2

Aidan Rowe's chapter takes as its point of departure the need for universities to respond to the current turbulent environment. Drawing on design-based learning, Rowe argues that design can offer unique and relevant teaching and learning practices to other disciplines that can help educators cross and blur disciplinary boundaries. Rowe, however, does not advocate design's primacy over other disciplines but instead places the emphasis on focusing on questions that matter. More specifically, Rowe's chapter focuses on critical stances, problem-focused learning, alternative communication methods, interdisciplinarity, collaboration, future-orientation, and human-centricity. Indeed, it is fascinating to witness how different disciplines are shifting their ethos towards human- or planet-centricity and we hope Rowe's chapter serves as food for thought for anyone interested in transforming their curricula from detached to up close.

Chapter 3

Brophy et al.'s chapter starts provocatively by urging us to consider the environmental and societal disasters taking place in the world right now. As of writing, the Intergovernmental Panel on Climate Change recently released their latest report (IPCC, 2022) and the key message is far from ambiguous: CO_2 emissions continue to increase, the richest 10% of households globally produce disproportionate amounts of greenhouse gas (GSG) emissions, as global warming continues to increase. As such, future graduates across educational institutions will end up working in an environment where everyone should feel the sense of urgency to act. Thus, it is no wonder that Brophy et al.'s chapter questions design's overt focus on creating innovations or serving capitalist gains. Drawing on their experiences in running a transdisciplinary design unit at Queensland University of Technology (Brisbane, Australia), Brophy et al. provide us with provocative and actionable insights on how we, as educators, might respond to the complex and tumultuous environment in which we engage with teaching and learning practices.

Chapter 4

Keane and Yeow, like the chapter of Rowe's and Brophy et al., also take the need to rethink higher education in the twenty-first century as their point of departure. Instead of going back to what has been argued for before, Keane and Yeow argue that Vargo and Lusch's (2004) Service-Dominant Logic (SDL) approach can provide us with novel ways to transform higher education. While SDL has been widely utilised in investigating and theorising how companies create and capture value together with their customers, to the best of our knowledge there is a paucity of studies utilising SDL in the context of higher education. As such, Keane and Yeow's chapter provides inspiring insights on how higher education institutions could collaborate with other actors (e.g. public and private sector) in creating research and teaching initiatives with capabilities to respond to contemporary and future challenges.

Chapter 5

Finally, Kelly's chapter is a beautiful treatment of the pedagogy of ambiguity in higher education. At least for us, the editors of this volume, ambiguity is simultaneously desirable and frightening, but Kelly's approach to ambiguity through sociolinguistics (e.g. Fairclough, 1993) was something we felt gave us two things: actionable concepts and theoretical understanding of how to deal with ambiguity. In essence, ambiguity is not only something we expose our students to, but it is also something we, as educators, should experience. Furthermore, Kelly discusses paradoxes and tensions with regard to ambiguities: what is ambiguous in our course designs, what is not? As such, dealing with ambiguity goes beyond tackling difficult problems or mitigating ambiguity; we also ought to discuss *how* we make sense of ambiguity, since taking matters at face value might do more harm than benefit us.

Chapter 6

In our first synthesis, Chew reflects on what the previous four chapters have in common, how they are situated *vis-à-vis* transdisciplinarity, and how we could transform our courses and programmes to better address societal and environmental issues, ambiguity, and notions of inclusion and exclusion. While we all acknowledge the need to respond to global challenges and, eventually, shift from reactivity to proactivity, at the same time this shift is a far cry from being a walk in the park, so to speak. As educators, we also have our own individual wants, fears, and desires, but how do we strike a balance between what the planet needs, what our students need, and what we need? Granted, as we are currently witnessing Russian army atrocities in Ukraine, it is inspiring to witness how people and organisations are willing to let go of their privileges to ensure a more just future for everyone. In essence, change is possible, and we are hopeful that the era of "sustainable growth" is coming to an end and we, humanity, are ready to compromise our hedonistic pleasures, driven by consumerism, to reconsider what is good for the planet for future generations.

Part II—Blending Boundaries: Design and Technology

Higher education institutions face a new operating environment as remote learning becomes more common, and the demand for it rises. Technology-mediated teaching and learning itself is by no means novel, but the high demand urges administrators, managers, and faculty alike to make remote learning a viable option for students. From design's point of view, remote learning poses both challenges and opportunities. While the virtual design studio has already been experimented on and researched in design education for a couple of decades (Jones et al., 2021; Lehtonen et al., 2021; Maher et al., 2000), other disciplines are still exploring how to effectively teach design online. Thus, we need a more nuanced and granulated understanding of design's signature pedagogies (as per Shulman, 2005) if we are to successfully continue utilising design as a transformative catalyst across disciplines, regardless of the mode of instruction.

As the Covid-19 pandemic has revealed, higher education institutions will need to assess their capabilities in technology-mediated teaching and learning. More specifically, exploring new technologies in design-driven teaching and learning needs to consider these two perspectives at least: first, how do we implement new technologies meaningfully and on what criteria are efficiency and meaningfulness grounded, and second, how do technologies feed back into our understanding of design pedagogies. In other words, how do we make a pedagogical case for new technologies without following fads and taking technology as an end in itself?

Chapter 7

In this chapter, Valencia et al. investigate how design education was impacted by the shift to online learning brought about by the Covid-19 pandemic. Instead of emulating the face-to-face classroom experience, Valencia et al.'s study focuses on learning styles and how learners can be supported by bringing the tactile dimension to the online classroom. In addition, since their study focuses on a course titled *Design, Entrepreneurship, and Innovation*, findings and insights from this chapter are also applicable to courses outside the design discipline. In effect, the tactile dimension in teaching and learning should not be understood to be exclusive to design and similar making-based disciplines, which is why Valencia et al.'s in-depth description of the course they taught provides food for thought for face-to-face classes, as well.

Chapter 8

As in the preceding chapter, Dickson-Deane et al. also takes the Covid-19 pandemic and how it nudged higher education institutions to rethink their approach to blended and online learning as its point of departure. More specifically, Dickson-Deane et al. focus on work-integrated learning, namely, how learning that takes place in the classroom can be effectively utilised in the workplace. This is a promising intersection since, as Dickson-Deane et al. posit, in the beginning of the pandemic online learning was aimed at responding to a global health emergency, but in order to move forward online learning needs to depart from emulating, towards creating meaningful learning experiences. As such, Dickson-Deane et al.'s thorough treatment of work-integrated learning from multiple perspectives (i.e. higher education, society, industry) provides us with insightful concepts and frameworks for blurring the boundary between the classroom and the surrounding society.

Chapter 9

Korenblat's chapter continues to blur the boundary between the classroom and the surrounding society by focusing on care and caring in design education. Drawing on feminist philosophers' work on the ethics of care, Korenblat illustrates how Eddy, a transformation design initiative at the State University of New York at New Paltz (SUNY New Paltz), holds considerable potential in instilling learners with a heightened sense of caring towards the environment and other individuals. More specifically, Korenblat mobilises Noddings's (1991) four elements of caring education: modelling, practice, dialogue, and confirmation. As such, this chapter provides a detailed case study of Eddy from the perspective of caring and ethics of care. While Korenblat's chapter focuses on design education, the ethics of care deserves attention across disciplines since by doing so, we can strike a balance between disciplinary knowledge and caring relations. In other words, education is not only about broadening one's horizon, but equal focus should be on how we relate to others and the surrounding environment.

Chapter 10

In the final chapter of Part II, Kristensen and Gabrielsen weave together marketing and design disciplines by questioning the concept of "market". While during the latter half of the twentieth century academic discussions at the intersection of design and marketing were quite vibrant, it is only recently that these conversations have been picked up once again. Here,

Kristensen and Gabrielsen's work has been pioneering at this intersection as they have continued exploring under what conditions design and marketing intersect (e.g. Kristensen & Gabrielsen, 2017). In this chapter, and through concrete examples, Kristensen and Gabrielsen discuss the conceptual differences and similarities between marketing and design. From a pedagogical perspective, such analytical accounts are necessary if we are to move beyond disciplines. Bringing disciplines together, and often going beyond them, requires that educators understand how their "home" discipline stands in relation to other disciplines in question.

Chapter 11

Derek Jones's brilliant synthesis brings together the chapters from a design educator's point of view, and as such, we believe this synthesis has merit for educators in and outside design discipline. First, for design educators, an explicit treatment of boundaries can equip us with words and methods to blur the boundaries, and second, for educators from other disciplines, understanding how designers and design educators approach boundaries (or implicitly treat them as something that *ought* to be blurred and questioned) helps in grasping blind spots with regards to boundaries in their own field. Furthermore, as Jones's synthesis engages in a dialogue with each of the chapters in Part II, we believe this conversation to offer additional vantage points to each chapter.

Part III—Capacities for the New World: Changing Roles and Responsibilities of Educators and Learners

Given that we consider the transformation of higher education institutions an ongoing and open-ended project, this leads us to acknowledge that learning takes place not only by or amongst students: educators and other stakeholders engage in learning, as well. While this claim is by no means novel, we think it is worthwhile to analyse and theorise on multiple stakeholder perspectives to better comprehend the changes and their implications taking place in the higher education sector. For example, during the last few years, there has been a rising tendency to incorporate design-driven methodologies into disciplines outside the design school context, but so far, the main arguments have been built upon design thinking as a stepwise model. This, then, begs us to ask, what kind of capacities are we nurturing in our students and how are signature pedagogies in other disciplines influenced and shaped by design? Moreover, what are the strategic and tactical goals driving the design agenda?

From an educator's point of view, transforming course and programme curricula often involves stepping out of comfort zones in an attempt to create something new. In this light, a future-oriented educator wears many hats, including that of a lecturer, information expert, learning experience designer, knowledge collaborator, media producer, facilitator, project manager, communication specialist, or coach. An active educator embodies the role of promoting a collaborative team culture and fruitfully challenging students but also supporting students to manage tasks, communicating about the process and progress, giving timely feedback, and enabling learners to move towards critical, yet creative, thinking. Knowledge expertise has traditionally been the most important asset of the educator, but facilitation expertise is an equally important asset in the modern learning environment. Thus, creating transformational learning experiences through curriculum development, more often than not, also implies changes and transformation in the educators involved in the process.

Chapter 12

Williams's chapter offers an insightful take on fashion education from the standpoint of cycles of action and reflection. More specifically, Williams argues that there are certain misalignments when it comes to cycles of fashion education, nature, society, industries, and culture. In this chapter, Williams takes us on a journey during which they followed fashion educators for over three years and the findings highlight the need for a systemic level transformation in fashion education. While the chapter provides a beautiful balance of theoretical and pedagogical insights, FashionSeeds (2022)—the platform that is referred to in this chapter—contains a plethora of actionable methods and tools that can be utilised not only in fashion education but in other disciplines, as well.

Chapter 13

Departing from fashion education in the context of the United Kingdom, Hoyos Bustamante's chapter describes how Universidad Nacional de Colombia (UNal) has engaged in cross-disciplinary course collaboration in the Colombian context to foster stronger connections between the university and its surrounding environment. One of the main motives for the collaboration has been to better equip the learners with skills relevant to the twenty-first century, and at the same time, there also seems to be an underlying desire for change at the institutional level. Furthermore, while Hoyos Bustamante approaches interdisciplinarity from architecture's point of view, reflections in this chapter go beyond single disciplines. In other words, focus is on how teaching and learning in a higher education institution could be transformed and with what consequences. As such, Hoyos Bustamante's work also broadens our body of knowledge by discussing interdisciplinarity in the Colombian context.

Chapter 14

While Hoyos Bustamante's chapter focused on the Colombian context, Yeo's chapter takes us almost to the other side of the world; namely, Singapore. Similar to Williams's chapter, Yeo also focuses on sustainability and sustainable futures, but from the perspective of the kind of skills future graduates might need in order to thrive in the future workplace. Singapore provides a fascinating context for such explorations since it is a country with few natural resources of its own. As a result, Singapore has become known as one of the leading countries globally when it comes to utilising design as a competitive advantage in both public and private sector. To this end, Yeo's chapter addresses the future of design education from the perspective of educators and learners, and this in-depth treatment can also be implicitly understood as an invitation for other disciplines to collaborate with design educators and practitioners. Most importantly, however, Yeo's chapter is a fascinating standpoint to understanding how design education might be developed to better understand future challenges and opportunities.

Chapter 15

In the last chapter in Part III, Inamura focuses on exploring how design has gone, and can go beyond human-centric needs. Acknowledging recent conversations in design that have called for more-than-human approaches to design practice and outcomes, Inamura also draws our attention to the notion that we still seem to be lacking a robust and diverse methodological foundation for post-human-centric design. To this end, he introduces meditative methods as a potential way forward. From a more concrete perspective, Inamura's chapter elucidates how design and engineering students utilised meditative practices and mindfulness going beyond human-centric design to propose planet-centric solutions to identified problems and challenges. Furthermore, and perhaps most importantly, Inamura's chapter, while focusing on design and engineering students, provides food for thought for educators across disciplines.

Chapter 16

Finally, Reymen et al.'s synthesis brings yet another geographical setting to Part III—the Netherlands. Especially interesting in this synthesis is Reymen et al.'s focus on changing responsibilities for educators and learners. Two points are worth mentioning here: first, the need to respond to societal and environmental issues is becoming increasingly relevant, and second, building on the previous point, learner ownership and empowerment will become critical issues.

In more concrete terms, such a call for transforming higher education also echoes with Lund Dean and Fornaciari's (Fornaciari & Lund Dean, 2013; Lund Dean & Fornaciari, 2013) reflections on how we might improve the course syllabus so we could see it not only as a contractual artefact but, more importantly, as a co-created tool that would help educators and learners in developing community-based inquiries. In other words, while there seems to be a wide-spread consensus on the need to transform higher education (institutions), we should also be careful so as not to fall into the *transformwashing* trap, which is why the message Reymen et al. are advocating should be taken seriously at all levels of higher education, ranging from the syllabus to the leadership level.

Looking Ahead

To conclude, we hope you will find this edited volume an inspiring reading experience. While the topics that many of the chapters are touching upon might be daunting, at the same time, we simply cannot argue for turning a blind eye to them. As such, this edited volume of contributions from across the globe will be released at a peculiar time. On the one hand, we are witnessing people coming together to reverse the trajectories set in motion by mass production and consumerism, and on the other hand, future scenarios have perhaps never been as ambiguous and volatile as they are today.

Nonetheless, we are hopeful. While some might discard this stance as being overtly naïve, at the same time the world also needs aspiration and forward-thinking ideas; whether it is about instilling our students with skills and competencies to tackle wicked problems, ensuring intergenerational justice not only in terms of people but all animals and the environment, creating commercially and environmentally successful innovations, or rethinking the role higher education institutions play in envisioning desirable futures, we all have the agency and the responsibility to act (e.g. hooks, 1994).

References

Boling, E., Schwier, R. A., Gray, C. M., Smith, K. M., & Campbell, K. (Eds.). (2016). *Studio teaching in higher education: Selected design cases*. Routledge.

Dar, S., Liu, H., Dy, A. M., & Brewis, D. N. (2021). The business school is racist: Act up! *Organization, 28*(4), 695–706.

Dewey, J. (1997). *Experience & Education*. Free Press.

Fairclough, N. (1993). *Discourse and social change*. Polity Press.

FashionSeeds. (2022). *FashionSeeds*. Retrieved April 14, 2022, from https://www.fashionseeds.org/

Fleming, P. (2020). Dark academia: Despair in the Neoliberal Business School. *Journal of Management Studies, 57*(6), 1305–1311.

Fleming, P. (2021). *Dark academia: How universities die*. Pluto Press.

Fornaciari, C. J., & Lund Dean, K. (2013). The 21st-century syllabus: From pedagogy to andragogy. *Journal of Management Education, 38*(5), 701–723.

Gass, J. R. (1972). Preface. In L. Apostel, G. Berger, A. Briggs, & G. Michaud (Eds.), *Interdisciplinarity: Problems of teaching and research in universities* (pp. 9–10). Organisation for Economic Co-operation and Development.

Harley, B., & Fleming, P. (2021). Not even trying to change the world: Why do elite management journals ignore the major problems facing humanity? *Journal of Applied Behavioral Science, 57*(2), 133–152.

Hill, R. (2012). *Whackademia: An insider's account of the troubled university*. University of New South Wales Press.

hooks, b. (1994). *Teaching to Transgress*. Routledge.

Intergovernmental Panel on Climate Change (IPCC). (2022). Climate change 2022: Mitigation of climate change. https://www.ipcc.ch/report/ar6/wg3/. Accessed online 5 Apr 2022.

Jantsch, E. (1972). Inter- and transdisciplinary university: A systems approach to education and innovation. *Higher Education, 1*(1), 7–37.

Jones, D., & Lotz, N. (2021). Design education: Teaching in crisis. *Design and Technology Education: An International Journal, 26*(4), 4–9.

Jones, D., Lotz, N., & Holden, D. (2021). A longitudinal study of virtual design studio (VDS) use in STEM distance design education. *International Journal of Technology and Design Education, 31*, 839–865.

Kismödi, E., & PItchforth, E. (2022). Sexual and reproductive health, rights and justice in the war against Ukraine 2022. *Sexual and Reproductive Health Matters, 30*(1), 2052459.

Kristensen, T., & Gabrielsen, G. (2017). Design economy—Value creation and profitability. In R. Cooper, S. Junginger, & T. Lockwood (Eds.), *The handbook of design management* (pp. 347–365). Bloomsbury.

Lehtonen, M. J., Khamisani, N., & Gatto, G. (2021). Playful absence/absence of play: Rethinking the design studio in online environments. *Design and Technology Education: An International Journal, 26*(4), 20–36.

Lehtonen, M. J., Yeow, P., & Chew, J.-Y. (2022). Empowering change for future-making: Developing agency by framing wicked problems through design. *Futures, 139*, 102952.

Lund Dean, K., & Fornaciari, C. J. (2013). The 21st-century syllabus: Tips for putting andragogy into practice. *Journal of Management Education, 38*(5), 724–732.

Maher, M. L., Simoff, S. J., & Cicognani, A. (2000). *Understanding virtual design studios*. Springer.

Mills, C. W. (1997). *The racial contract*. Cornell University Press.

Mitchell, C. (2022). Widening access to HE starts in early childhood. https://wonkhe.com/blogs/widening-access-to-he-starts-in-early-childhood/. Accessed online 9 May 2022.

Noddings, N. (1991). *The challenge to care in schools: Rethinking the goals of education*. Teachers College Press.

Piaget, J. (1972). The epistemology of interdisciplinary relationships. *Interdisciplinarity: Problems of teaching and research in universities* (pp. 127–139). Organization for Economic Cooperation and Development.

Shulman, L. S. (2005). Signature pedagogies in the professions. *Daedalus, 134*(3), 52–59.

Vargo, S. L., & Lusch, R. F. (2004). Evolving to a new dominant logic for marketing. *Journal of Marketing, 68*(1), 1–17.

PART I

Designing Design Education: Strategies for Navigating Complexities and Boundaries

CHAPTER 2

Design Pedagogy: Higher Education Possibilities for the Twenty-First Century

Aidan Rowe

Introduction

Higher education in the twenty-first century faces unique challenges, changing contexts, and opportunities (Birdsall, 1999; Tapscott, 2012). To respond to and address these issues academia has looked to incorporate a variety of new learning and teaching practices. Some common approaches that have been proposed include attempts to: increase interdisciplinary learning opportunities to address education across disciplines (Bear & Skorton, 2019; Klaassen, 2018); create increased collaborative learning situations enabling students to work with a wider range of people (Helfand, 2014) and; ensure students have experiential curricular learning experiences to ground and extend their education in meaningful situational contexts (University of Toronto, 2017).

A common factor in these identified approaches is that these learning practices—and others—have a long history and are commonly used within

A. Rowe (✉)
University of Alberta, Edmonton, AB, Canada
e-mail: aidan.rowe@ualberta.ca

M. J. Lehtonen et al. (eds.), *Design Education Across Disciplines*,
https://doi.org/10.1007/978-3-031-23152-0_2

design education. In this chapter, I argue that design pedagogy, particularly framed through what Davis (1998) terms Design-Based Learning (DBL), offers experience with unique teaching and learning practices to other, non-design areas in academia, and can serve as an inspirational educational model for the twenty-first century.

Design-Based Learning commonly employs a range of exciting, interconnected, robust, and needed learning and teaching practices that include:

- Students assuming a critical stance questioning existing practices and then responding;
- Learning through problem-focused scenarios;
- Employing a variety of alternative communication methods in the studio;
- Interdisciplinary and interprofessional learning opportunities where students work with other fields;
- Collaborative learning situations where students work with—not just for—other actors;
- Situating design as a future-oriented activity;
- Curricular and co-curricular experiential learning experiences that position education beyond the classroom;
- A focus on hands-on creation where students learn through doing; and,
- Framing design activity through a human-centred approach.

While common to Design-Based Learning these pedagogical practices are often at the cutting edge of other academic disciplines. This chapter begins by situating the author and the work and then articulating a broad overview of the changing space and needs of twenty-first-century higher education, it then describes how learning takes place within the design studio (the central location for Design-Based Learning). The paper then identifies and contextualizes nine characteristics found in Design-Based Learning. It concludes by recognizing the areas within design education that need further development.

It is hoped that by noting the unique practices and methods utilized in design education we are, first, able to recognize, reaffirm, build upon and further incorporate them into our pedagogical practices. Secondly, it also enables opportunities for a critique of DBL, recognizing strengths and weaknesses in this model. And finally, it creates prospects to articulate the possibilities of applying them externally to other academic areas.

Importantly this interrogation allows us to ask questions about what and who we teach, but also larger questions of why we teach. If we conceptualize higher education in a broad sense—and after Simon's (1969) thoughts on design—as the "changing of existing situations into preferred ones" we must engage with the edges and the possibilities of our pedagogies (p. 130).

Situating the Author and the Work

Educated in both Canada and the UK, the author has taught design full-time for over 20 years in Canada, England, and Germany. Originally situated in visual communication design he has taught across a wide range of areas including industrial design, fashion design, service design, design for health, and visual communication design (and spaces in between). He has held positions in art and design colleges (UK), a technical university (Germany), and a large, research-intensive public university (Canada).

Increasingly he has worked in, and with, other academic areas integrating DBL practices in non-design academic areas including nursing, medicine, engineering, and public health. Importantly, this work has helped to inform and extend his teaching practice in design.

The possibilities discussed here—identifying practices found within design pedagogy that could benefit other academic areas—are grounded in his own experience. While these possibilities do not present a unified framework that is applicable to all areas, they do represent opportunities for interrogating design education and improving higher education pedagogies.

Changing Needs and Demands in Higher Education

Universities and colleges today face complex challenges, evolving contexts, and dynamic opportunities. To prepare for these challenges and to respond to them, a variety of learning and teaching requirements have been identified (and, in some instances, have been implemented).

The call for new pedagogical practices to address these needs is being driven by a range of factors including student demand, shifting industry wants, growing recognition of wicked problems, pedagogical advancements, and broader societal and cultural shifts and requirements (Robertson, 2021; University of Toronto, 2017). What is broadly recognized though is that the traditional, siloed university learning model that is representative of much of contemporary higher education is outdated and ineffective at meeting the needs of the twenty-first century (Robertson, 2021; Thomas & Brown, 2009).

This call for change is not new though, the formative educational theorist John Dewey (1963) placed great importance on the learner's experience and active participation in the learning process (Schubert, 1996). More recently, in the 1960s, Canadian media theorist Marshall McLuhan (with Fiore, 2001) questioned the traditional, top–down specialist pedagogy model declaring "[e]ducation must shift from instruction … to discovery—to probing and exploration" (p. 100). In 1972, the polymath Erich Jantsch noted the growing complexity of the world, stating that there is a need for "a type of education which fosters judgement in complex and dynamically changing situations" (pp. 101–102). More recent pedagogic research supports this broader shift.

For example, recent work has (re)enforced the (re)declaration of creativity, innovation, and critical thinking as crucial skills for twenty-first-century students, abilities needed to prepare for a future that will be characterized by continually shifting demands, relationships, ecologies, and workplaces (Rampersad & Patel, 2014). Egan et al. (2017) describe these skills as "key to effective learning in higher education and beyond" (p. 21). McLaughlan and Lodge (2019) reinforce this claim, noting that graduates entering "professions in the twenty-first century will require an enhanced capacity for innovation and adaption to change" (p. 1). Thomas and Brown (2009) declare that the one defining feature of this century is that it will be "characterized by constant change" and that these skills are foundational for future students (p. 1).

The need for experiential learning has also been recognized as a crucial pedagogical opportunity for students. Experiential learning—falling under numerous names including co-ops, practicums, work experience, community-service-learning, work-integrated learning, amongst others—creates genuine learning opportunities for students—often outside the classroom—that locates and broadens their studies in meaningful environments (Schubert, 1996). Students can apply their learning in new, external situations while also building skills to work meaningfully with partners and collaborators. Eyler (2009) states that this "helps students both to bridge classroom study and life in the world and to transform inert knowledge into knowledge-in-use". And while forms of experiential learning are more common in vocational or professionally oriented programmes they are not as pervasive in more traditional areas of the university system (Eyler, 2009).

Growing opportunities for students to learn between and across disciplinary and functional boundaries have also been identified as a crucial area that needs further development (Jamieson et al., 2022). These opportunities enable students to "understand and make connections across a diverse array of knowledge and skills, they embark on a path to more rewarding lives and employment opportunities" (Bear & Skorton, 2019, p. 60). Relatedly, students need the further possibility of learning and collaborating with a range of partners and collaborators (in and out of university settings) that cross professions, histories, and futures. These collaborative pedagogical opportunities help to put learning "into a situated context that deals with systems and identity as well as the transmission of knowledge" (Thomas & Brown, 2009, p. 1).

In addition to working across boundaries and collaborating with a range of partners, students need meaningful opportunities to consider issues surrounding globalization and culture (Rowe, 2013). In a continually connected world that crosses borders and continents, educators need to structure learning that addresses broader conceptualizations of culture and globalization including asking questions of longstanding inequities and continued imbalances. As Robertson (2021) asks, how "we might insert the idea of the global and the development of global competences, into teaching and learning" (p. 2).

Learning focused on expanded and extended forms of communication is also needed today. Traditional academic settings often rely upon specific and established forms of communication models—often replicating the instructor's own learning experiences (Canniffe, 2018). Frequently these are formal, traditional, and summative, and while these forms are still important there is also a need to expand the opportunity for students to gain a broader range of communication skills that embrace different technologies, audiences, futures, and needs, so that students are, as Parker (2009) notes "encouraged to think more laterally about the sites and spaces in which those skills could be used" (p. 15).

There has also been an increasing focus on recognizing the growing scale and complexity of social system problems faced in the twenty-first century. These problems span disciplines, are ill-formulated and pernicious, and there are no simple solutions. Often these are termed "wicked problems" (Rittel & Webber, 1973) and they can serve as a big idea curricular framework in higher education. The growing use of the United Nations Sustainable Development Goals (2015) in education settings offers similar opportunities, where students address issues that do not sit within established disciplinary boundaries (e.g. climate action, gender inequality, etc.) and that require collaboration and innovative forms of thinking and doing.

These identified the twenty-first-century educational needs, while not an exhaustive list, help to articulate the changing spaces of higher education. Additionally, they document the need to develop and implement innovative learning practices to help address these challenges and embrace the opportunities presented. As Thomas and Brown (2009) note, for educational institutions to take advantage of these opportunities the pedagogical response needs "to be as rich and complex as the challenges and opportunities we face" (p. 15).

The Design Studio: A Brief Overview

Design education, particularly that which is practiced in a studio setting, offers a range of unique pedagogical practices. Shulman (2005) defines the shared pedagogical practices found in the design studio as a form of "signature pedagogy" in that they are "pervasive and routine, cutting across topics and courses, programs and institutions" (p. 56).

The studio setting usually involves a smaller cohort learning environment, typically between 12 and 20 students, that meet for longer periods (e.g. six hours per week, sometimes broken up into smaller bi-weekly blocks) than traditional university courses. The studio blends problem and inquiry-based learning using a "cognitive apprenticeship model" with a focus on design-based responses to identified challenges, briefs, and problems (McLaughlan & Lodge, 2019, p. 4). Importantly, briefs are broad and often loosely defined, ensuring students have an opportunity for further research and refinement.

Students may work formally in groups on briefs, or individually, and the studio is a highly collaborative environment with near-continual discussion, observation, feedback, and contribution. Feedback may be delivered individually (instructor to student) and during class, or, through a public (within the class) presentation called a critique (crit). Often, within a crit an instructor may take the lead in providing feedback, students may be called to introduce their work, and additionally, they may be required to critique the work of fellow students. There is a focus on constructive feedback, where comments are actionable and address areas to improve. Importantly, feedback given at these stages is often formative, enabling opportunity for reflection, refinement, and improvement through iterative development. This situating of constructive failure as a central component of the studio is fairly unique in comparison to the majority of higher education environments, where "failure is viewed negatively and curricula are specifically designed to mitigate the risk of student failure" (McLaughlan & Lodge, 2019, p. 5).

Characteristics of Design-Based Learning

Design-Based Learning offers a range of unique and specific practices that collectively present design pedagogy as a distinctive educational experience within the university. Davis (1998, p. 7) articulates the benefits of DBL particularly in regard to current challenges, noting that in DBL we "find dynamic examples of learning and problem-solving perfectly suited to an environment of ever-expanding information, diverse citizen needs, and great uncertainty created by the failure of traditional problem-solving models". Building on the notion of signature pedagogies, Shulman (2005, p. 54) notes the collective nature of these prevalent and unifying practices (across programmes, schools, and even countries) and that they "implicitly define what counts as knowledge in a field and how things become known". While many of these individual characteristics may be found in the learning practices of other academic disciplines—for example, nursing, computer science, or engineering—the collective nature presents a unique and important educational exemplar. Importantly, many of the teaching and learning practices located within Design-Based Learning address identified needs of students, the modern university, and society.

The characteristics listed above are neither exhaustive nor compulsory nor are they as individual as noted, they often weave together, supporting and extending one another. Different programmes, schools, or sub-disciplines within design might focus on some more than others, just as others may be more tangential. In some form or other, they appear in Design-Based Learning.

Central to design education is the requirement for students to assume a critical stance, where they question existing practices and then respond to them. This criticality is applied to both existing situations and artefacts

(Why was this done? Could we do this?) in addition to their work and the work of fellow students (Does this work? How could this be better?). Postman and Weingartner (1969) refer to this as developing the anthropological perspective in students, so they can both exist in a culture and separate themselves from it to critique it. McLaughlan and Lodge (2019) describe the process of students applying criticality to their work (and that of classmates) as a form of Socratic dialogue, helping them to identify and externalize opportunities and weaknesses of their work. This iterative process of making, assessing, reflecting, and making again is central to design pedagogy.

Learning through problem-focused scenarios facilitates opportunities for students to play an active role in creating alternate futures in response to identified situations. Historically—and, too often currently—design is framed as a problem-solving discipline, where a specific problem is presented to a designer (or design students) and they respond, presenting their solution. The shift from problem-solving to problem-focused (also called problem-based, problem-finding, or problem-seeking) moves the process from a passive (receiving the brief) to an active (creating the brief) learning environment, where students have the opportunity to better understand "the very constraints and conflicting requirements that make problems difficult" (McLaughlan & Lodge, 2019, p. 7). Importantly, Marenko and Brassett (2015) note that this shift enables the design process to focus on invention rather than attempting to reveal solutions that may already exist.

Design-Based Learning also employs a variety of communication and presentation methods. While learning in the studio employs many traditional academic modes of teaching—e.g. lectures, presentations, etc.—it also employs a range of unique practices that blur the line between instructor and student. These include crits, where instructors and students constructively critique the designed artefact (whether product, system, or service), this is often a formative exercise where there is a later opportunity for students to refine their work. As students work together in the studio on the same project there are official and unofficial forms of co-learning taking place at all times, from the casual observation of a fellow student's work, direct discussion of the brief, and asking for feedback and guidance. As Shulman (2005, p. 54) notes "[s]tudents are experimenting and collaborating, building things and commenting on each other's work without the mediation of an instructor". The instructor is also circulating around the studio checking in individually with students and providing personalized guidance and feedback. While the instructor is clearly in charge of the class, learning and teaching are taking place at many levels and from many sources. Shraiky and Lamb (2013, p. 467) describe this rich environment as offering a mix of "problem-based, action-based and practice-based learning".

Design is a multidisciplinary profession and design education fosters this through interdisciplinary and interprofessional learning opportunities where students work with other disciplines and fields in a variety of roles (Cheatham, 2017). There is a growing recognition of the need for genuine interdisciplinary experience, particularly driven by the growing complexity of societal issues that design has the opportunity of addressing (Friedman, 2012). These interdisciplinary opportunities take a variety of forms within design education including pathways or streams allowing design students to work with students in other areas (e.g. engineering, computer science), to courses designed around thematic challenges (i.e. big ideas), or external briefs where students might work with a collaborating partner. Design education also enables opportunities for interprofessional learning where students from different disciplines model collaboration and learning within an educational setting helping to establish good working practices for later professional life (Buring et al., 2009).

Collaborative learning situations are a key feature of Design-Based Learning, where students work with—not just for—other actors (fellow students, users, collaborators, etc.) throughout the design brief. These collaborative learning scenarios ensure that students are interactive, and importantly, as Shulman (2005) notes, create extended forms of co-accountability, where students are responsible to their peers and not just their instructors. Davis (1998, p. 9) notes that design is a social, and not an individual, activity, and that responsibility for its outcome is shared "with the audiences who make meaning of it through its use". The range of collaborators is also important as students gain an opportunity for negotiation, relationship building, and the modelling of future roles helping to prepare them for life after graduation. At full realization, this collaboration occurs throughout the whole of the learning process from problem identification, to joint work on the iterations, to students partaking in forms of assessment, both formative and summative, Shraiky and Lamb (2013, p. 462) note that as most evaluative processes are public in DBL, "students become participants in the evolution and improvement of each other's work".

Design—and by extension design education—is a future-oriented activity, one that imagines and brings to life that which does not yet exist, and as Ward (2015, p. 229) notes this allows us to "push the boundaries of knowledge", design has a crucial role—and opportunity—to play in bringing social, political, and environmental change to life, and it is not only concerned with the "creation and materialisation of possible worlds, but also a way of thinking and critically responding to current issues and concerns" (Marenko & Brassett, 2015, p. 4). Designers occupy a space between what is today and what will be tomorrow. In a time of nearly constant change, where the world seems to be continually speeding up, the ability to both imagine new futures and bring them to life is a critical and needed skill.

Curricular and co-curricular experiential learning experiences that situate education beyond the traditional classroom (e.g. client-based briefs, community-engaged learning, etc.) have long been a feature of Design-Based Learning. While these learning opportunities fall under a variety of names—from co-ops to work-integrated learning to practicums—design education incorporates these learning experiences within the overall curriculum enabling learners to interact with the world and, importantly, to integrate "new learning into old constructs" (Eyler, 2009, p. 24). Importantly, there has been growing recognition within higher education of the power and possibility of these learning opportunities, for example, a University of Toronto (2017, p. 2) white paper noted that growing its experiential learning opportunities would "collectively enhance both the student learning experience and the University's ability to support broader community and societal needs".

Central to Design-Based Learning is a focus on learning through hands-on creation and making. The process of making is a thread running through an entire project, with various stages of creation taking place, and this iterative process—where a designer makes anew and improves upon previous versions—is central to studio activity. It needs to be emphasized that making is a form of learning, a way to understand and construct knowledge, and not just a final expression of that learning (Davis, 1998). We see the value of the experience of making also articulated in educational theories around constructionism where it is argued that knowledge is constructed through real-life experiments that enable learning and "pairing abstract concepts with concrete experiences to make sense of knowledge" (Loh, 2018, p. 139). Ward (2015, p. 229) powerfully describes this when he notes the unique role that design has in the academy, that we "make things, to make sense of the world".

We have also seen recognition of the power and intrigue of making and hands-on creating taking place through the popularization of makerspaces—sometimes called hackerspaces, fab labs, or garages. These spaces are often embedded within educational settings—sometimes in a library, otherwise freestanding—giving access to a variety of tools and systems (Pendergast, 2020). Benefits and possibilities include increased engagement in learning, addressing issues of belonging and equity, identity development, and opportunities for students to learn how to learn (Nadelson, 2021).

Finally, design—as a discipline and pedagogy—is most successful when framed through a human-centred approach where there are genuine opportunities for designing with those we design for. Historically, design that looks to involve end-users in the process has fallen under many titles—participatory design, co-design, user-centred design, etc.—but all recognize that working with—and, not just for—other humans throughout the design process helps to ensure the validity, appropriateness, and possibility of design responses (Buchanan, 2004; Noël, 2017).

As noted, the characteristics identified above are neither exhaustive nor required within Design-Based Learning, but this list captures much of the teaching and learning activities that take place in design pedagogy. Identifying its key practices creates an opportunity for educators to interrogate these methods, celebrating their characteristics and possibilities but also allowing the opportunity for critique, revision, and improvement.

Work to Be Done: A Concise List

It is important to note that there are many areas within design education—as in any academic discipline—that require interrogation and improvement (Frascara, 2017; Friedman, 2012). For example, design pedagogy often suffers from a focus on the aesthetics and form rather than the outcomes—how well it works—of the products, systems, and services designed. Cheatham (2017, p. 76) notes that design education models that "focus primarily on form-giving are too ideologically narrow or practically inflexible to address" the complexity of challenges today. Thiessen (2017, p. 148) also notes the need for design to shift away from what she terms "prioritizing the object" to a broader consideration of design processes and outcomes.

Related, there needs to be further work establishing more rigour in research methods and practices in design education. As discussed in previous work (Rowe, 2020, p. 51), design is a fairly recent addition to academia, and as such it often lacks its own "formal, established research frameworks and theoretical practices". Often, within current design education, research involves a superficial visual collection of existing practice—a reinforcement of Thiessen's "prioritizing the object" (2017, p. 148); there needs to be more work establishing solid research practices, possibly looking to other disciplines and fields—for example, psychology, anthropology, and ethnography—for models.

A third area needing progress is aligning design education with a more culturally and socially responsible practice. Design—as professional practice—has a long history as an artefact-based, consumer-focused, reactive profession and work was often lacking broader considerations of cultural, social, and ecological impact (Margolin & Margolin, 2002; Papanek, 2006). Linked to, and building upon the previously identified needs, further interrogating design education with regard to its cultural and social effects and possibilities is crucial (Burns et al., 2006). There is also a critical need to build upon more recent and important work exploring decolonizing design (for example, the work of Dori Tunstall). Furthering this, Irwin (2016, p. 91) argues that design and its powerful approach to addressing problems can "serve as a catalyst for positive social and environmental change".

Continued work in these areas—and there are others—is needed in design education today, and while there are exemplar programmes and individuals, further effort is needed. As Noël (2020, p. 6) notes the knowledge, models, and exemplars exist to strengthen design pedagogy, what is needed is "a solid implementation strategy to make the change real".

Conclusion

There have been consistent calls for change in higher education teaching and learning practices to address a world of "infinite complexity, endless possibility, and near constant change" (Thomas & Brown, 2009, p. 15). These demands come from many areas including faculty, industry, society, governments, and most importantly, students. Identified responses include a variety of new and enhanced teaching and learning practices including creating learning environments that embrace collaboration and interdisciplinarity; embedding rich experiential learning opportunities within the curriculum; and, employing a problem-based learning paradigm to encourage active, engaged learning (Bear & Skorton, 2019; Canniffe, 2018; University of Toronto, 2017).

While many of these pedagogic practices are more recent to the majority of academia, they—and other needed practices—have a long history in design pedagogy, particularly as described by Davis (1998) as Design-Based Learning. I have argued here that DBL offers experience with these teaching and learning practices to other areas of the university.

Key pedagogical practices that are found in Design-Based Learning include learning situations framed through a critical stance; employing problem-focused design scenarios; using a range of alternative communication methods in the studio; embedding interdisciplinary and interprofessional learning opportunities often within collaborative and curricular and co-curricular experiential learning situations; and framing design as a hands-on, future-oriented, and human-centred activity.

Naming these practices allows us—as design educators—to further investigate how and where we employ them in our teaching and learning practices. Importantly they also enable opportunities for critique and refinement, recognizing both the possibilities and limitations offered. These practices from Design-Based Learning are not a panacea to address the growing complexity of the twenty-first century, but they do create the prospect for other academic areas and disciplines to look to practices with DBL as inspiration and evidence of effective learning strategies to help educate students to thrive and lead in the twenty-first century.

As Thomas and Brown (2009, p. 15) state there is a stark need to interrogate not only what and who we teach but also why, and that "our approach to education and learning needs to be as rich and complex as the challenges and opportunities we face". Design-Based Learning offers a history of "curricular innovation, proven pedagogy, and student achievement" that places design pedagogy at the centre of higher education's needed transformations (Davis, 1998, p. 13).

References

Bear, A., & Skorton, D. (2019). The world needs students with interdisciplinary education. *Issues in Science and Technology, 35*(2), 60–62.

Birdsall, N. (1999, September 14). *Education, globalization and demands of the 21st century*. Carnegie Endowment for International Peace. https://carnegieendowment.org/1999/09/14/education-globalization-and-demands-of-21st-century-pub-44

Buchanan, R. (2004). Human-centered design: Changing perspectives on design education in the East and West. *Design Issues, 20*(1), 30–39.

Buring, S. M., Bhushan, A., Brazeau, G., Conway, S., Hansen, L., & Westberg, S. (2009). Keys to successful implementation of interprofessional education: Learning location, faculty development, and curricular themes. *American Journal of Pharmaceutical Education, 73*(4), 60.

Burns, C., Cottam, H., Vanstone, C., & Winhall, J. (2006). *Transformation design*. Design Council. www.designcouncil.org.uk/red

Canniffe, B. J. (2018). US design education prepares for careers in the past, is standing on the edge of the abyss, looking away and singing kumbayah. *Graphic Design Journal, 7*, 7.

Cheatham, D. M. (2017). A multiple intelligences model for design: Developing the ways designers think as design disciplines expand. *Dialectic, 1*(2), 75–100.

Davis, M. (1998). Making a case for design-based learning. *Arts Education Policy Review, 100*(2), 7–15.

Dewey, J. (1963). *Experience and education: The Kappa Delta Pi Lecture Series* (First Collier Books Edition). Collier Book.

Egan, A., Maguire, R., Christophers, L., & Rooney, B. (2017). Developing creativity in higher education for 21st century learners: A protocol for a scoping review. *International Journal of Educational Research, 82*, 21–27.

Eyler, J. (2009). The Power of Experiential Education. *Liberal Education, 95*(4), 24–31.

Frascara, J. (2017). Design, and design education: How can they get together? *Art, Design & Communication in Higher Education, 16*(1), 125–131.

Friedman, K. (2012). Models of design: Envisioning a future design education. *Visible Language, 46*(1), 132–153.

Helfand, D. (2014, February 5). We evolve, but the university stands still. *The Globe and Mail*. http://www.theglobeandmail.com/news/national/education/we-evolve-but-the-university-stands-still/article16689680/

Irwin, T. (2016). Redesigning a design program: How Carnegie Mellon University is developing a design curricula for the 21st century. *Solutions, 6*. https://thesolutionsjournal.com/2016/02/22/redesigning-a-design-program-how-carnegie-mellon-university-is-developing-a-design-curricula-for-the-21st-century/

Jamieson, M. V., Ead, A. S., Rowe, A., Miller-Young, J., & Carey, J. P. (2022). Design at scale in a first-year transdisciplinary engineering design course. *International Journal of Engineering Education, 38*(1), 14–24.

Klaassen, R. G. (2018). Interdisciplinary education: A case study. *European Journal of Engineering Education, 43*(6), 842–859.

Loh, P. (2018). Making as pedagogy: Engaging technology in design teaching. In N. Llevot-Calvet & O. B. Cavero (Eds.), *Advanced learning and teaching environments—Innovation, contents and methods* (pp. 137–155). InTech.

Marenko, B., & Brassett, J. (eds.). (2015). *Deleuze and design*. Edinburgh University Press.

Margolin, V., & Margolin, S. (2002). A "social model" of design: Issues of practice and research. *Design Issues, 18*(4), 24–30.

McLaughlan, R., & Lodge, J. M. (2019). Facilitating epistemic fluency through design thinking: A strategy for the broader application of studio pedagogy within higher education. *Teaching in Higher Education, 24*(1), 81–97.

McLuhan, M., & Fiore, Q. (2001). *The medium is the massage: An inventory of effects*. Gingko Press.

Nadelson, L. S. (2021). Makerspaces for rethinking teaching and learning in K–12 education: Introduction to research on makerspaces in K–12 education special issue. *The Journal of Educational Research, 114*(2), 105–107.

Noël, G. (2017). Health design: Mapping current situations, envisioning next steps. *The Design Journal, 20*(sup1), S2304–S2314.

Noël, G. (2020). We all want high-quality design education: But what might that mean? *She Ji, 6*(1), 5–12.

Papanek, V. (2006). *Design for the real world: Human ecology and social change*. Thames & Hudson.

Parker, S. (2009). *Social animals: Tomorrow's designers in today's world*. RSA Design & Society.

Pendergast, S. C. (2020). *Makerspaces in higher education: Student engagement*. University of Calgary.

Postman, N., & Weingartner, C. (1969). *Teaching as a subversive activity*. Dell Publishing.

Rampersad, G., & Patel, F. (2014). Creativity as a desirable graduate attribute: Implications for curriculum design and employability. *Asia-Pacific Journal of Cooperative Education, 15*(1), 1–11.

Rittel, H. W., & Webber, M. M. (1973). Dilemmas in a general theory of planning. *Policy Sciences, 4*, 155–169.

Robertson, S. L. (2021). Global competences and 21st century higher education—And why they matter. *International Journal of Chinese Education, 10*(1), 221258682110103.

Rowe, A. (2013). Design pedagogy and cultural interrogation: Considering the international and the anthropological. *Design Principles and Practices an International Journal: Annual Review, 6*, 12.

Rowe, A. (2020). Participatory action research and design pedagogy: Perspectives for design education. *Art, Design & Communication in Higher Education, 19*(1), 51–64.

Schubert, W. H. (1996). Perspectives on four curriculum traditions. *Educational Horizons, 74*(4), 169–176.

Shraiky, J. R., & Lamb, G. (2013). Studio-based learning in interprofessional education. *Journal of Interprofessional Care, 27*(6), 461–468.

Shulman, L. S. (2005). Signature pedagogies in the professions. *Daedalus, 134*(3), 52–59.

Simon, H. A. (1969). *The sciences of the artificial*. MIT Press.

Tapscott, D. (2012, October 15). Discovery learning is the new higher learning. *The Globe and Mail*. http://www.theglobeandmail.com/news/national/time-to-lead/discovery-learning-is-the-new-higher-learning/article4610656/

Thiessen, M. (2017). I don't know, I just like it: Exploring how design students think about criticism. *Art, Design & Communication in Higher Education, 16*(2), 145–156.

Thomas, D., & Brown, J. S. (2009, June). *Learning for a world of constant change: Homo Sapiens, Homo Faber & Homo Ludens revisited*. 7th Glion Colloquium, Los Angeles, University of Southern California.

United Nations. (2015). *THE 17 GOALS|Sustainable development*. https://sdgs.un.org/goals

University of Toronto. (2017). *Rethinking higher education curricula: Increasing impact through experiential, work-integrated*. University of Toronto.

Ward, M. (2015). Rapid prototyping politics: Design and the de-material turn. In W. Jonas, S. Zerwas, & K. von Anshelm (Eds.), *Transformation design: Perspectives on a new design attitude* (pp. 227–245). Birkhauser.

CHAPTER 3

Socially Responsive Design Education: Emerging Designers and Authentic Transdisciplinary Collaborations

Claire Brophy, Deanna Meth, Melanie Finger, and Dean Brough

> "…we are living in a world of massive and rapid change. We must be alert to new developments and prepare our students for a changing world-not only in technology but in the needs and expectations of the human beings whom we ultimately must serve." Richard Buchanan, World Congress, International Council of Graphic Design Associations, 1998.[1]

[1] Even now, more than twenty years on, Buchanan's words still ring true. We refer the reader to his more recent writings for further discussion (Buchanan, 2019).

C. Brophy (✉) · D. Meth · M. Finger · D. Brough
Queensland University of Technology (QUT), Kelvin Grove, Brisbane, QLD, Australia
e-mail: cj.brophy@qut.edu.au

M. J. Lehtonen et al. (eds.), *Design Education Across Disciplines*,
https://doi.org/10.1007/978-3-031-23152-0_3

Introduction

There is little denying that our complex global system is in the midst of unprecedented change—from bushfires to social unrest, to health-care systems on the verge of collapse. The sustainability of our system relies on carefully balanced interactions between several layers—culture, governance, infrastructure, commerce, fashion/art, and of course, nature (Brand, 2018). When an imbalance exists between the layers, results can be catastrophic (Brand, 2018). These are *wicked problems,* and complex societal challenges that are almost impossible to solve (Buchanan, 1992; Rittel & Webber, 1973, cited in Skaburskis, 2008). Working to understand them though, and envision preferable futures (Candy, 2010) is what design does best. And pushing the practice to be more socially responsible is something design writers and critics have been imploring designers to do for decades (Buchanan, 1992, 1998; Margolin, 1998; Mau, 2020). But while the complexity of global problems—and the opportunities for design—continues to increase, challenges remain in transforming the practice.

One challenge with this shift in perspective for design is that since the industrial revolution design has been perceived as an industrial or trade profession (Buchanan, 1998), concerned with the making of things. Despite progressive, and radical visions from early schools like the Bauhaus, almost as soon as formalized education for design emerged, students were funnelled into conventional courses related to trade professions and constrained by disciplinary boundaries, or as Blevis (2016) describes, 'guild logic'. This conventional model persists today (Max-Neef, 2005).

Another consequence of such conventional approaches is that design has a fascination with progress (Escobar-Tello et al., 2021) that grew from the Western, Eurocentric visions of the Bauhaus itself. Often preoccupied

D. Meth
e-mail: deanna.meth@qut.edu.au

M. Finger
e-mail: m.finger@qut.edu.au

D. Brough
e-mail: d.brough@qut.edu.au

with innovation, differentiation, and a culture of consumption (Margolin, 1998), the impact of this fascination is evident in every layer of our complex system. There are growing calls for a decolonization of design,[2] though while design practice continues to mature, it is still typically disconnected from the social, cultural, and environmental consequences of the work (Kolko, 2012; Vodeb, 2015).

Before the turn of the last century, while writing on developments in environmentally conscious, sustainable approaches to design, Victor Margolin (1998) said that the rhetoric of design discourse was generally at odds with reality and at that time he called for the design to rethink its role in the world. Around the same time, while discussing developments such as the advance of design thinking, Richard Buchanan (1998) was reflecting on the relationship between design practice and design education. Instead of education following behind practice, as it had traditionally, Buchanan envisioned a deepening relationship between industry and academia to anticipate and explore new conditions of practice. While illustrating powerful developments in what she defines as *Attitudinal Design*, Alice Rawsthorn (2020) takes it back even earlier. Rawsthorn draws on the post-war visions of Moholy-Nagy and the idea that design should be '...transformed from a specialist function into an attitude of resourcefulness and inventiveness...' where projects were motivated by the major issues of the time, and their '...relationship with the need of the individual and the community' (Moholy-Nagy, 1947, p. 42 in Rawsthorn, 2020, p. 7). While design practice is going through transformations, despite these early visions, even now very little of design education includes the *critical discourses* necessary to understand the 'agency of design' (Vodeb, 2015, p. 426).

How then might design education motivate emerging designers to turn their practice to the indeterminate nature of wicked problems in our global system (Buchanan, 1992, 2019; Kolko, 2012). We argue that this push is not required—it is expected. And that to live up to these expectations and enable future designers to take on a more socially responsive role in the world, transdisciplinary design education is crucial. We interpret transdisciplinary education as including substantive knowledge fundamental to disciplines, as well as other ways of knowing (McGregor, 2017).

[2] Emerging practices are as diverse as the Indigenous contexts in which they are situated. We refer the reader to Schultz, et al. (n.d.). What is at stake with decolonizing design? A roundtable. *Design and Culture*, 10(1), 81–101.

Four features characterize transdisciplinary learning: (1) related to socially relevant issues, (2) transcending disciplinary paradigms, (3) participatory research with those experiencing complex social issues, and (4) a search for 'unity of knowledge' (Pohl, 2011, cited in McGregor, 2017). With transdisciplinary ways of knowing, students learn to 'co-create, co-disseminate and co-use transdisciplinary knowledge, which emerges from the iterative interactions between disciplines and the rest of the world' (McGregor, 2017, p. 3).

In this chapter, we illustrate the pedagogic model for a first-year transdisciplinary design unit at Queensland University of Technology (QUT) in Brisbane, Australia, called Impact Lab 2: People. As a direct response to institutional 'Blueprint 6' ambitions to provide real-world learning (QUT, 2019). The teaching leverages fundamental principles and methods of design and applies them to social objectives. By leaning heavily on design theory and adopting an outward-facing participatory mindset, the unit takes students out of their disciplinary silos, and advocates for the potential of transdisciplinary design practice grounded in notions of the social role of design.

Impact Lab 2 is partially motivated by what Peruccio et al. (2019), call a *consciousness of complexity*—a growing and shared awareness of how interconnected our societal systems are. Healthcare, transport, and human services—often designed for outdated models of society, these systems do not come without fault. Discussions of such faults have recently shifted from the expert to the public domain (Peruccio et al., 2019, p. 18). People want to change for ecological, economic, cultural, or societal reasons, or any combination of these and more—fuelling people's *consciousness* is an expectation of change. For all its iterations, design has always been an agent of change (Rawsthorn, 2020). Much of this push for better practice is coming from younger generations (Kolko, 2012); evidenced by growing participation in global climate strikes and political protests, an increasing number of young people expect improvements in politics, in their workplaces and educational institutions, including greater ethical practices and engagement in the process of change from higher education in particular.

Background: Design Education Contexts and Purposes

The Changing Role of Higher Education

When writing, fifty years ago, about the purpose and role of universities, Jantsch (1970, p. 9) suggested they were generally determined by a long-standing 'profound belief in a *laissez-faire* type of self-organization of science and technology'. Based on conventional theories of science (which at the time were rather disconnected from any social activities or goals), universities were oriented towards 'a rigid model of society, with well-defined and unchanging patterns of professional specialization' (Jantsch, 1970, p. 9). Staff and faculty of the time had begun calling for changes to perceived elitism in higher education; however, bureaucratic fixation with 'past cultural patterns and states of society' was a main factor in resistance to change (Jantsch, 1970, p. 10).

Universities continue to be criticized for being privileged, outdated institutions (Lee, 2021), slow or unable to adapt and respond to global crises. In July 2020, in an online conference looking at the effect of the Covid-19 pandemic on higher education, Arizona State University President Michael Crow (2020) criticized the rigid faculty and discipline models as oversimplistic and inadequate for our complex, interconnected society, saying such models limited the ways in which disciplines and universities evolved. To avoid continuing to drive negative ecological outcomes and social inequities, Crow (2020) said that universities must adapt to new forms of intellectual engagement—rather than being the agent that grounds society, institutions must begin to empower society and consider, '...broader sets of intellectual enterprise, transdisciplinary, post-disciplinary, interdisciplinary, as well as disciplines themselves...'. Compounding these kinds of challenges are decades-old conflicts and competing interpretations on the role of higher education (Solbrekke & Sugrue, 2020).

Solbrekke and Sugrue (2020) present conventional interpretations of higher education as an instrumental-educational polarity—at one end, narrow, skill-oriented interpretations, at the other, wider notions that lean more towards *genuine education* and critical reflective practice. Similar to Crow's 'broader sets of intellectual enterprise' (2020) Solbrekke and Sugrue (2020) argue for ambitious approaches to higher education driven by notions of public good. To do this, they say, higher education must be located in the wider world, occupying 'a discursive space, of give and take, of reason and reasoning, of care and sensibility, of unity and difference, of equal participation across members of society…' (ibid., p. ix).

Similarly, Jantsch (1970, p. 12 said that the purpose of universities must reflect their role in enhancing society's capability for continuous self-renewal. Emphasizing pluralism, social objectives, long-range outcomes, and positive leadership, Jantsch (1970) proposed changes to the primary functions of universities—education, research, and service—as well as a multi-level systems approach to educational policy. Jantsch's education/innovation system defines varying types of disciplinarity (e.g. multi, inter, etc.); transdisciplinarity, the ultimate aim, happens when the whole system is working towards a defined common purpose. This type of values-based, ecosystem perspective (Schultz, n.d.) is the foundation for transdisciplinary learning (McGregor, 2017) and more recently transdisciplinary design (Blevis, 2016).

Shifting Perspectives in Design Practice and Education

While once predominantly a trade profession, concerned with tangible products or the built environment, design continues to mature and evolve. Developments such as design thinking (Buchanan, 1992), have pushed design practice, reluctantly some may say, into organizational strategy and business innovation. A discussion on the varying views of this approach to design practice is outside the scope of our writing here, though in any case, it would be difficult to dispute the elevated profile brought on by such developments. Increasingly, design is being called-on to actively participate in the process of social change (Mau, 2020)—this is a role that makes sense. Complex social systems were designed, and many, such as our health and welfare systems now need to be re-designed, or at the very least, critiqued and better understood. Design provides ways to understand and describe the present, as well as methods to imagine, illustrate, and move towards alternative futures (Candy, 2010).

Kolko (2012) suggests that the negative consequences of complex societal challenges or wicked problems (Rittel & Webber, 1973, cited in Skaburskis, 2008), can be 'mitigated through the process of design' (p. 11). Indeed Rawsthorn's (2020) account of the work of attitudinal designers is testament to the powerful potential of the practice. Given the generally accessible and creative methods used in design, such emerging approaches aimed at social innovation or public good, are being adopted by government departments and organizations worldwide (Australian Design Council, 2021; Victorian Government, 2021). Between 2009 and 2016, the UK Design Council (2018) estimates that the UK 'design economy' grew by 52%, with much of that coming from designers working in what would be considered non-traditional design industries. The recently re-established Australian Design Council (2021) with a mission of championing the role of design in addressing 'complex, social, economic and environmental challenges' clearly demonstrates this evolution of design practice from products to services and systems. For a wide range of reasons, most of which reflect the breadth of influences on, and expectations of higher education (Meth et al., 2021), design schools have however varied in their responses to shift from traditional design practice towards reimagining transdisciplinary design education.

Design theory speaks of design as a whole (Cross, 2011; Dorst, 2015). The reality, however, is that the design process in traditional design education (Evans et al., 2021) is rarely this holistic. We observe, for example, students welcomed into a collective'School of Design' only to be filtered into seven disciplinary silos, with little opportunity for ***common learning*** (Müller et al., 2005). 'Uni-disciplinary' is still the dominant educational model for most universities, particularly at undergraduate levels (Max-Neef, 2005). Disciplines compete for prominence and 'academic prestige' (Max-Neef, 2005), while administrative structures and funding stifle collaboration. From a design perspective, Blevis (2016) describes this as 'guild logic', where 'disciplinary boundaries are defended in order to control who is credentialed as a matter of membership' (2016, p. 5). While there is a need for specialized practice, re-designing complex societal systems requires far more knowledge and skill than can be offered by any single discipline; it demands collaboration (Blevis, 2016; Kolko, 2012; Vodeb, 2015).

Grounded in the signature pedagogies of design and informed by characteristics of transdisciplinarity (Blevis, 2016; Jantsch, 1970), and transdisciplinary learning (McGregor, 2017; Pohl, 2011), we use the case of our first-year transdisciplinary design unit, Impact Lab 2: People, to illustrate a range of techniques and concepts that might foster socially responsive design education.

Case Study in Practice: Impact Lab 2: People

Impact Lab 2: People is part of a suite of four units of study spanning QUTs three-year Bachelor of Design degree programme (Place, People, Planet and Purpose, Fig. 3.1). Part of a bold reimagining of the design degree in 2019 (Meth et al., 2021) the unit includes upwards of 800 students from seven disciplines: industrial design, interaction design, visual communication, fashion design, interior design, architecture, and landscape architecture. A strength of Impact Lab 2 is its use of collaborative community-led projects that place transdisciplinary design pedagogy within real-world contexts—what Vodeb (2015) describes as extra-disciplinary, socially responsive design.

Through authentic collaborative projects with external industry and community organizations, the model for Impact Lab 2 brings to the fore the capacity, and in fact the responsibility for design to engage in complex social issues. Students choose from project briefs designed in collaboration with a breadth of external partners—from large, national charities, to small, local social enterprises. Working in groups, students are encouraged to adopt a participatory, designer-as-mediator mindset

Fig. 3.1 The suite of four transdisciplinary Impact Labs woven through the Bachelor of Design degree

(see Manzini, 2014; Sanders, 2002) to address issues such as homelessness, mental health, energy poverty, and domestic violence. Students are encouraged to acknowledge the impact of small changes on big systemic issues and consider their own agency in such actions through leveraging systems thinking to break down complexity and design-led methods to envision preferable futures (Candy, 2010).

Below we outline the techniques supporting design theory and design students' conceptual groundings across themes such as culture, diversity, and inclusion (Jones, 2007; Noel, 2020; Sparke, 2004; Vink & Koskela-Huotari, 2020).

Considering Collaborative, Inclusive, Adaptable, and Responsive Design

As mentioned earlier, Impact Lab 2 turns first-year design students towards the opportunity and responsibility (Manzini, 2014; Margolin, 1998) for using purposeful design to enact meaningful social change. Adopting a socially constructed, learning by doing approach (see Dewey, 1938), the content includes underpinning theoretical thinking, examples of that theory in context, and design tools and activities that demonstrate that way of thinking in practice. To help locate design in the *wider world* (Solbrekke & Sugrue, 2020), the unit leverages the UK Design Council's Systemic Design Framework (2020) as an underlying guide. The model emphasizes the *invisible activities* surrounding the design process, including—among others: considering the history, societal values, and assumptions of the system; spending time with communities on their terms; creating inclusive spaces; and using a variety of engagement methods (Design Council UK, 2020).

Lectures discuss the unique, complex, interconnected characteristics of societal challenges (Kolko, 2012). Contemporary examples (e.g. racism, equality, health, and loneliness), are framed in the context of Wicked Problems (Kolko, 2012; Rittel & Webber, 1973, cited in Skaburskis, 2008), and relevant, familiar exemplars that speak to many disciplines are provided such The Spiky Blob (Antonelli & Rawsthorn, 2020) and The Teeter-Totter Wall (Rael San Fratello, 2021) are provided. The examples illustrate that while critical societal challenges span boundaries and cultures (Sparke, 2004), they are not understood or experienced by all people in the same way.

The design process, and differences between expert-discipline and transdisciplinary design mindsets, are discussed early in the semester. Disciplinary knowledge related to human factors, usability and User-centred design, and design goals such as effectiveness, efficiency, and performance, are presented and critiqued with reference to seminal works such as Measure of Man (& Woman) (Tilley Alvin et al., 2002) and the Design of Everyday Things (Norman, 2013). The focus then shifts to Inclusive Design[3] and the importance of including culture and values in design as ways to shift the focus 'beyond usefulness' (Gilmore et al., 2008). As well as using Hofstede's Cultural Model (Jones, 2007) we refer to the work of Penny Sparke (2004) and the notion that all kinds of cultural values make their way into the designed artefact. Again, several contemporary examples from products to systems are included to illustrate the theory in context (e.g. the MeToo movement; the cultural issues within the Australian Parliament [Tiernan, 2021] and, under the shadow of racial unrest in the United States, the re-launch of bandages in diverse skin tones [Klein, 2020]). The theoretical content concludes with a discussion on sense-making through frame creation (Dorst, 2015) as an important way to refine the complexity and ambiguity of the social contexts explored during the design process.

Tools to Demonstrate Theory in Action

Motivated by social objectives (Jantsch, 1970) and notions of public good (Solbrekke & Sugrue, 2020), Impact Lab 2 aims to illustrate the social role of design (Vodeb, 2015), to uncouple the 'problem-solving' power frequently bestowed upon design practice (Blevis, 2016; Kolko, 2012), and shift the perceived elitism of universities (Jantsch, 1970). To do this, students are introduced to notions of *designing with and* for (Kolko, 2012) and approaches such as participatory design (Kensing & Greenbaum, 2012) and Co-Design (McKercher, 2020).

Participatory or co-design techniques support signature pedagogies of design education and transdisciplinary learning (McGregor, 2017). Participatory approaches level the playing field (Kensing & Greenbaum, 2012), they are motivated by consensus through collaboration, and the

[3] Inclusive design celebrates diversity and emphasizes the limitations of 'one-size-fits-all' design approaches. The UK Design Council (2006) provides a wonderful overview on the principles of Inclusive Design.

creation of a space where the knowledge and practices of diverse stakeholders are considered (Buur & Matthews, 2008; Kensing & Greenbaum, 2012). The nature of Impact Lab 2 means that it is not possible to completely realize these approaches in our projects; as first-years, our students are at the very beginning of their design practice, and there are ethical implications with Co-Design projects focused on sensitive topics such as domestic violence, for example. However, the guiding principles and models from Participatory and Co-Design—including but not limited to, power sharing, mutual learning, and active partnerships,—provide an important, and teachable foundation for purposeful transdisciplinary design work. Another key characteristic of participatory techniques is their '...interventionist methods of engagement' (Buur & Matthews, 2008, p. 7), and our studio activities rely heavily upon the principles of these techniques.

The Design Studio as a Collaborative Space

Studios are energetic spaces; dedicated to practice—to exploring the tools without fear of failure, and to experiencing working collaboratively in a way that transcends disciplines. Activities are collaborative and make use of many tangible materials, whiteboards, post-it notes, concrete walls and chalk, paper worksheets, Lego, and online platforms such as Padlet and Miro. Tools are introduced in a scaffolded approach, and collectively, they become a workbook to guide groups of students through the design process during the community-led project. Studio activities introduce students to the thinking behind various tools, including those used in design and design research. Using Sanders' (2002) *Say, Do, Make* model, the differences in these are discussed alongside the benefits of visual, tangible, and experiential techniques as ways to reframe problems by facilitating participation by diverse groups of people.

Positionality and Diversity in the Design Process

By drawing out socio-cultural similarities and differences, the Positionality Wheel (Noel, 2021) is a tool that makes diversity explicit. When used with our whole cohort as a transdisciplinary design team (Fig. 3.2), the visual tool makes the strengths and weaknesses of the team explicit, and helps facilitate discussions on possible harmful assumptions, biases, and misconceptions in design practice (Noel, 2021).

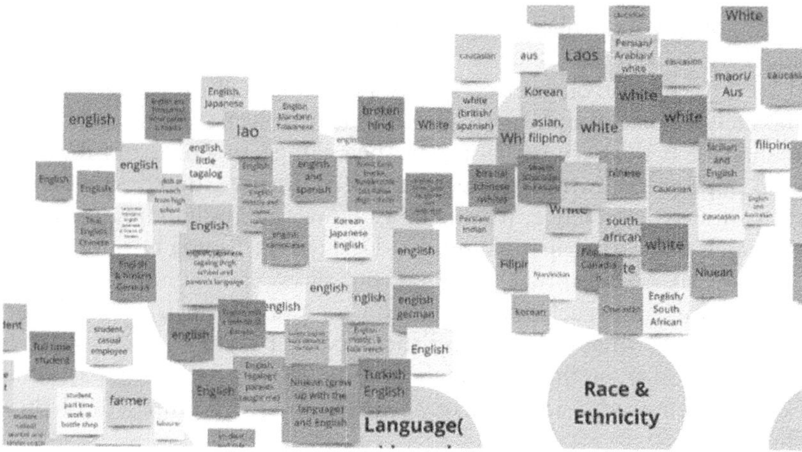

Fig. 3.2 An excerpt from a collective visual of the strength and weakness in our design team using the Positionality Wheel in Miro (Noel, 2021)

When completed in groups, other activities such as Draw a Bike (Lawson, 2006) and How to Make Toast (Wujec, 2021; Fig. 3.3) highlight diversity and illustrate the complex nature of even the most ordinary, everyday things. By visualizing mental models, these tools also demonstrate that sketching can be a useful method for helping break down complexity and working towards a shared vision (Wujec, 2021; Young, 2008). In our studios, the 'how to make toast' models made the personal preferences or geographic influences of our students explicit—(vegemite versus peanut butter, for example, or fridge storage versus cupboard storage).

Most models that the students built illustrated the physical product or outcome while overlooking people as part of the system. Reflecting the reality of collaborative design in diverse groups, consensus was not always possible or preferred, in fact, some students found developing a shared model of 'how to make toast' more challenging than expected. These simple group activities provide a foundation for discussion on more sensitive social systems. If trying to accurately describe how a bicycle works is challenging and the results across a group of people so varied, how might we navigate complexities and intricacies when designing, say, a government's welfare system?

Fig. 3.3 Student group work on trying to reach a consensus on a model of how to make toast

Considering Complexity, Wider Systems, and Processes

Framed by the notions of divergent, convergent, and systems thinking (Buchanan, 2019) (Double diamond), tools such as concept mapping (Kolko, 2010, 2012), insight statements, and problem definition (Nesta, 2021) are also introduced as ways to understand complexity. From a design perspective, these tools also help groups of students explore the context, expand the design brief, and identify more opportunities for meaningful design outcomes. By physically making, or prototyping, tangible modelling (Buur, 2018) takes the teaching and learning one step further.

Speaking to Sanders' *Participatory Mindset*, her *Say, Do, Make* model (2002), as well as the prototyping phase in design, tangible modelling is a Participatory Design technique that uses novel, dynamic physical artefacts to represent components of a system and their relationship to one another (Mitchell & Buur, 2010, p. 29; Fig. 3.4).

Using curated kits of objects, in our case Lego, students work in groups to map out and build representations of the system, context, or concept they are seeking to understand (Fig. 3.5).

These activities are playful and accessible—decisions need to be explained and justified, and complex language gives way to everyday, natural descriptions. As well as being accessible, the technique facilitates abstraction and the development of metaphors which stimulate meaningful discussions (Buur, 2018; Mitchell & Buur, 2010) and support the necessary, and often missing practice of reflection in design (Kolko, 2010, 2012; Vodeb, 2015).

Fig. 3.4 Say, Do, Make' model (Sanders, 2002) and students' drawings of bikes (Lawson, 2006)

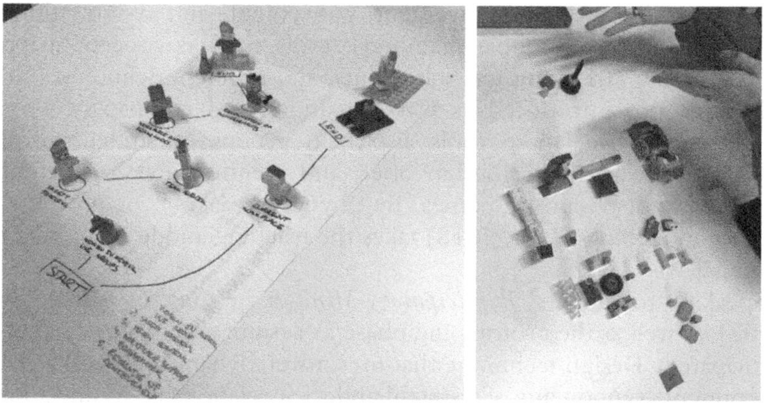

Fig. 3.5 Student groups using Lego for Tangible Modelling (Mitchell & Burr, 2010) to visualise the context and system (left) and prototype the concept (right)

Reflective Practice and Engendering a Critical Discourse

In our design unit, reflective activities and assessments provide opportunities for students to consider and interpret their learnings from the concepts, techniques, and approaches presented and discussed in the lectures and studios. Students are encouraged to consider the impact of what they have experienced on their perspective of design and their future practice as a designer; less about defining design, and more about what it means to them (Sonneveld & Hekkert, 2008). The work is completed individually and shared with the cohort (Fig. 3.6). Importantly, in reflective practice the context of Impact Lab 2 requires students to consider aspects such as community, culture, power, values, privilege, and exclusion, and the role that these play, and need to play in design.

Reflective practice provides the critical discourse (Vodeb, 2015) necessary for design students to realize the social role of design (Sonneveld & Hekkert, 2008), as well as the space necessary to begin transforming conventional uni-disciplinary thinking (Max-Neef, 2005). The pedagogical model of Impact Lab 2 breaks the boundaries of guild logic (Blevis, 2016) to enable more purposeful, values-driven (Jantsch, 1970) design. The 'extra-disciplinary' approach (Vodeb, 2015) provided through the collaborative, externally led projects further helps bridge gaps between practice and the academy, and opens students up to the genuine opportunity and value for design applied in this way.

> *'The unit taught me that there are multiple ways to confront complex global issues through design, which helps broaden my perspective on design in terms of problem solving.' Student feedback 2021.*
>
> *'I thought design was all about how a thing looked, I never thought there would be so many layers and things to keep in consideration. I also never thought that design could actually affect people negative and positively, I just saw design as an object not a process.' Student feedback 2021.*

Fig. 3.6 Sharing student reflections with the whole cohort (left) and individual reflections (right)

As part of ongoing action research that examines the impacts of our curricular intentions and our own critical reflective practice on the design of the unit, feedback is sought from students throughout and at the completion of the programme. While the above may seem an ambitious approach (Solbrekke & Sugrue, 2020) to design education, the methods are supported through evidence of the positive impacts such learning has had on students' development as both designers and humans.

Discussion and Final Thoughts

Just as design thinking turned design towards strategic and organizational opportunities, collaborative, transdisciplinary, and experiential design education can turn emerging designers towards the role of design as an agent of change (Rawsthorn, 2020). Grounded in fundamental design theories and methods, the Impact Lab 2: People model offers a way to motivate students to design for social objectives (Jantsch, 1970) and public good (Solbrekke & Sugrue, 2020). With roots in visionary post-war interpretations, this is more than an evolution of design thinking, this is a coming of age for design practice.

Underpinned by theory related to transdisciplinary thinking, the unit leverages participatory techniques to translate the theory into practice. By demonstrating systems thinking and reflection-in-action, the learnings engender a critical discourse and emphasize the importance of positionality, culture, and diversity in design processes focused on social change.

Mitigating the negative consequences of complex societal challenges demands collaborative practice (Kolko, 2012); so too does the reimagining of design education. Though we have illustrated a scalable model for purposeful transdisciplinary design education—its realization depends on institutions committed to radically re-thinking simplistic and inadequate discipline and faculty models (Crow, 2020); outward-facing institutions with a clear view of purpose (Jantsch, 1970), prepared to foster deeper collaborative relationships within and beyond the academy. For our degree programme, and the Impact Labs in particular, this transformation has been previously captured by Meth et al. (2021).

Fifty years ago, Jantsch envisioned a shift to a more dynamic, flexible structure in universities where students, academics and entrepreneurs had freedom to determine the path of their research and education. The university would be oriented towards society's values; actively shaping the future by initiating and maintaining close connections and information flow with government, industry, and other institutions.

Rawsthorn's (2020) recent account of attitudinal designers details this purposeful shift in organizations and individuals. Empowered by new digital technologies, crowdfunding, grants, and motivated sectors keen on overhauling outdated methodologies, Rawsthorn says designers are seizing 'the opportunity to pursue their political, cultural, and ecological concerns by operating independently' (2020, p. 11). As envisioned by some (Buchanan, 1998; Margolin, 1998) design practice has begun to embrace other fields (Rawsthorn, 2020), working in a way that transcends disciplinary boundaries to realize positive societal impact. Unlike Buchanan's (1998) vision though, design education is falling further behind practice. By continuing to follow conventional ideas of design, concerned with innovation and differentiation (Kolko, 2012; Margolin, 1998) design education can maintain the status quo, bound by 'guild logic' (Blevis, 2016) and the making of things. On the other hand, outward-facing, socially responsive institutions can confer agency upon emerging designers for social change and the critical transdisciplinary work ahead.

References

Antonelli, P. & Rawsthorn, A. (October, 2020). Alissa Eckert on designing the 'spiky blob' Covid-19 medical illustration. *Wallpaper: Guest editor issue.* https://www.wallpaper.com/design/october-2020-issue-free-download

Australian Design Council (2021). https://australiandesigncouncil.org/

Blevis, E. (2016). Must do, want to do, can do, can know (transdisciplinary design). *Design 3.0 forum.* http://design3-0.org/2016/wp-content/uploads/2016/10/MustWantCanKnow-Design3.0-Position-Blevis-V1.0.6d.pdf

Brand, S. (2018). Pace layering: How complex systems learn and keep learning. *Journal of Design and Science.*

Buchanan, R. (1992). Wicked problems in design thinking. *Design Issues, 8*(2), 5–21.

Buchanan, R. (1998). Education and professional practice in design. *Design Issues, 14*(2), 63–66.

Buchanan, R. (2019). Systems thinking and design thinking: The search for principles in the world we are making. *She Ji, 5*(2), 85–104.

Buur, J. (2018). Tangible business interviews. In P. V. Freytag & L. Young (Eds.), *Collaborative research design* (pp. 175–194). Springer.

Buur, J., & Matthews, B. (2008). Participatory innovation. *International Journal of Innovation Management, 12*(03), 255–273.

Candy, S. (2010). *The futures of everyday life: Politics and the design of experiential scenarios*. University of Hawai'i at Manoa

Cross, N. (2011). *Design thinking: Understanding how designers think and work*. Berg.

Crow, M. (2020, July 13). *REMOTE fireside chat with ASU President Michael Crow [video] YouTube*, https://www.youtube.com/watch?v=XD8CI1jwZf4&t=311s

Design Council UK. (2006). *The principles of inclusive design. (They include you.)*. https://www.designcouncil.org.uk/sites/default/files/asset/document/the-principles-of-inclusive-design.pdf

Design Council UK. (2018). *The design economy 2018*. https://www.designcouncil.org.uk/resources/report/design-economy-2018

Design Council UK. (2020). *Beyond net zero: A systemic design approach*. https://www.designcouncil.org.uk/resources/guide/beyond-net-zero-systemic-design-approach

Dewey, J. (1938). *Education and experience*. Simon and Schuster.

Dorst, K. (2015). *Frame innovation: Create new thinking by design*. The MIT Press.

Escobar-Tello, M. C., Ruette-Orihuela, K., Gough, K. V., Fayad-Sierra, J. A., & Velez-Torres, I. (2021). Decolonising design in peacebuilding contexts. *Design Studies, 73*, 101001.

Evans, R., Kelly, N., & Kerr, J. (2021). Being a design academic: Design process, practice, thinking. *How to be a design academic* (pp. 13–28). CRC Press.

Falconieri Hays. (2020). How I built a 3-D model of the coronavirus. *Scientific American*. https://www.scientificamerican.com/article/how-i-built-a-3-d-model-of-the-coronavirus-for-scientific-american/

Gedenryd, H. (1998). How designers work: Making sense of authentic cognitive activities. *Lund University Cognitive Studies, 75*, 1–123.

Gilmore, D. J., Cockton, G., Churchill, E., Kujala, S., Henderson, A., & Hammontree, M. (2008). *Values, value and worth: their relationship to hci?* Paper presented at the CHI'08 extended abstracts on human factors in computing systems.

IPCC. (2021). *Climate change widespread, rapid and intensifying*. IPCC Press Release, 9 August 2021 (2021/17/PR). https://www.ipcc.ch/site/assets/uploads/2021/08/IPCC_WGI-AR6-Press-Release_en.pdf

Jantsch, E. (1970). Inter- and transdisciplinary university: A systems approach to education and innovation. *Policy Sciences, 1*, 403–428.

Jones, M. L. (2007). *Hofstede-culturally questionable?* Paper presented at the Oxford Business & Economics Conference, Oxford, UK. Retrieved from: https://ro.uow.edu.au/cgi/viewcontent.cgi?article=1389&context=commpapers

Kensing, F., & Greenbaum, J. 2012. Heritage: Having a say. In Routledge *International handbook of participatory design* (pp. 41–56). Routledge.

Klein, K. (2020). Band-Aid launches bandages to "embrace the beauty of diverse skin". *Dezeen*. https://www.dezeen.com/2020/06/16/band-aid-bandages-brown-black-skin-tones/

Kolko, J. (2010). Abductive thinking and sensemaking: The drivers of design synthesis. *Design Issues, 26*(1), 15–28.

Kolko, J. (2012). *Wicked problems: Problems worth solving*. Ac4d.

Lawson, R. (2006). The science of cycology: Failures to understand how everyday objects work. *Memory & Cognition, 34*(8), 1667–1675.

Lee, B. (2021). *Who gets to be smart?* Allen & Unwin.

Manzini, E. (2014). Making things happen: Social innovation and design. *Design Issues, 30*(1), 57–66.

Margolin, V. (1998). Design for a sustainable world. *Design Issues, 14*(2), 83–92.

Mau, B. (2020). *Bruce Mau-MC24: Bruce Mau's 24 principles for designing massive change in your life and work* (p. 2020). Phaidon Press Limited.

Max-Neef, M. A. (2005). Foundations of transdisciplinarity. *Ecological economics, 53*(1), 5–16.

McGregor, S. L. (2017). Transdisciplinary pedagogy in higher education: Transdisciplinary learning, learning cycles and habits of minds. In P. Gibbs (Ed.), *Transdisciplinary higher education: A theoretical basis revealed in practice* (pp. 3–16). Springer.

McKercher, K. A. (2020). *Beyond sticky notes: Doing co-design for real book*. Beyond Sticky Notes.

Meth, D., Thomson, S., & Brough, D. (2021). Design curricula: Navigating process and people. In A. Blackler & E. Miller (Eds.), *How to be a design academic* (pp. 245–268). CRC Press. https://doi.org/10.1201/9780429351693-21

Mitchell, R., & Buur, J. (2010, August). Tangible business model sketches to support participatory innovation. In *Proceedings of the 1st DESIRE Network Conference on creativity and innovation in design* (pp. 29–33).

Moholy-Nagy, L. (1947). *Vision in motion*. Paul Theobald.

Müller, D., Tjallingii, S., & Canters, K. (2005). A transdisciplinary learning approach to foster convergence of design, science and deliberation in urban and regional planning. *Systems Research, 22*(3), 193–208.

Nesta. (2021). *Problem definition*. https://media.nesta.org.uk/documents/Resources_2017_version_10.pdf

Noel, L.A. (2020, June 4). *Envisioning a pluriversal design education*. [Conference paper]. DRS2020, PIVOT, https://drs2020.org/

Noel, L. A. (2021). Learning to recognize exclusion. *Journal of Usability Studies, 16*(2), 63–72.

Norman, D. (2013). *The design of everyday things: Revised and expanded edition.* Basic books.

Peruccio, P., Menzardi, P., & Vrenna, M. (2019). Transdisciplinary knowledge: A systemic approach to design education. In *Proceedings of 'DRS Learn X Design 2019: Insider Knowledge', Fifth International Conference, 9–12 July 2019, Ankara, Turkey* (pp. 17–23).

Pohl, C. (2011). What is progress in transdisciplinary research? *Futures, 43*(6), 618–626.

QUT (Queensland University of Technology). (2019). *Blueprint 6.* https://www.qut.edu.au/about/blueprint-for-the-future

Rael San Fratello. (2021). *Teeter-Totter Wall.* https://www.rael-sanfratello.com/made/teetertotter-wall

Rawsthorn, A. (2020). *Design as an Attitude.* JRP Ringier.

Readfearn, G. (2021, August 9). What does the IPCC's report mean for Australia, and what can we expect in the future?. *The Guardian.* http://www.theguardian.com/environment/2021/aug/09/what-does-the-ipccs-report-mean-for-australia-and-what-can-we-expect-in-the-future

Rittel, H. J., & Webber, M. M. (1973). Dilemmas in the general theory of planning. *Policy Science, 4*, 155–169.

Sanders, E. B. N. (2002). From user-centered to participatory design approaches. In *Design and the social sciences* (pp. 18–25). CRC Press.

Sanders, E., & Dandavate, U. (1999). Design for experiencing: new tools. In *First International Conference on Design and Emotion, TU Delft.*

Schultz. (n.d.). What is ecosystemology? https://nature.berkeley.edu/classes/ecosystemology/what%20is%20ecosystemology.pdf

Skaburskis, A. (2008). The origin of "wicked problems." *Planning Theory & Practice, 9*(2), 277–280.

Solbrekke, T. D., & Sugrue, C. (Eds.). (2020). *Leading higher education as and for public good: Rekindling education as praxis.* Routledge.

Sonneveld, M., & Hekkert, P. (2008). Reflection on being a Designer. In *DS 46: Proceedings of E&PDE 2008, the 10th International Conference on Engineering and Product Design Education*, Barcelona, Spain, 04–05.09.2008 (pp. 529–534).

Sparke, P. (2004). *An introduction to design and culture: 1900 to the present* (2nd ed.). Routledge.

Tiernan, A. (2021). *Yes, the culture in Parliament House is appalling. But there are systemic problems that also need urgent reform.* https://theconversation.com/yes-the-culture-in-parliament-house-is-appalling-but-there-are-systemic-problems-that-also-need-urgent-reform-155755

Tilley, A. R., Dreyfuss, H., & Wilcox, S. B. (2002). *The measure of man and woman: human factors in design. Revised Edition.* John Wiley & Sons.

Victorian Government (2021). *Human-centred design playbook*. From https://www.vic.gov.au/human-centred-design-playbook

Vink, J., & Koskela-Huotari, K. (2020). Unpacking the nature of social structures as the materials of service design. [Conference paper]. *ServDes.2020 Tensions, Paradoxes and Plurality, Melbourne, Australia*. https://ep.liu.se/konferensnummer.aspx?series=ecp&issue=173

Vodeb, O. (2015). Social innovation and design education: Towards a socially responsive communication design pedagogy. *Design and Culture, 7*(3), 423–431. https://doi.org/10.1080/17547075.2015.1096187

Wujec, T. (2021). *Draw how to make toast*. From https://www.drawtoast.com/

Young, I. (2008). *Mental models: aligning design strategy with human behavior*. Rosenfeld Media.

CHAPTER 4

Blurring the Boundaries: The New Collaborative Education, Work and Skills Ecosystem

Kate Keane and *Pamela Yeow*

Introduction

What do problem-solving, effective communication and critical thinking have in common? All three appear in a list of top ten essential skills identified by wide-ranging websites and think tanks, from the World Economic Forum Future of Jobs Report (2020) to YouGov polls. It is clear that in considering future skills and the future of work, the way we design (or re-design) teaching and learning in higher education needs to be

K. Keane (✉)
Canterbury Christchurch University Business School, N Holmes Road, Canterbury CT1 1QU, UK
e-mail: kate.keane@canterbury.ac.uk

P. Yeow
School of Business, Economics and Informatics, University of London, Birkbeck Malet Street, Bloomsbury, London WC1E 7HX, UK
e-mail: P.yeow@bbk.ac.uk

M. J. Lehtonen et al. (eds.), *Design Education Across Disciplines*,
https://doi.org/10.1007/978-3-031-23152-0_4

transformed so that future graduates' capabilities, skills and knowledge provide a relevant combination of both technical expertise and intangible knowledge and skills. In embracing an interdisciplinary and collaborative approach, this chapter proposes that by contextualising technical skills learning within complex problem statements such as those made by the United Nations Sustainable Development Goals (SDG) (United Nations Department of Economic and Social Affairs, n.d.), students will be able to couple real-life learning with real-life solutions and thereby develop a sense of purpose.

In this chapter, we put forward the use of Service-Dominant (S-D) Logic (Lusch & Vargo, 2006) to reimagine the relationship between Higher Education Institutions (HEI) and their students. Using one of the foundational premises of S-D logic, we propose that an open, collaborative environment where genuine value co-creation can take place can benefit all stakeholders within the HE ecosystem to meet the challenges of Industry 4.0. S-D logic has developed through research over the last 20 years on goods-service exchange in post-industrial economies and offers a framework, congruent with the shift from the manufacturing of goods to service product dominance that identifies value co-creation between actors in the activity of service exchange. We propose that in an ecosystem of shared values based around knowledge exchange, learning and creation, all stakeholders (students, educators, researchers, employers, industries, wider society and government) can provide input and derive value, and all members can simultaneously give and receive resources and knowledge.

Successful case studies will illustrate how a systems-level approach (Senge, 1990) that encourages students to think collaboratively, across subject silos and work with industry experts alongside their academic tutors is the way forward towards developing students and graduates that are fit for the future and the future of work. We conclude with an ecosystem approach to enable higher education educators to engage in value co-creation to meet the needs of the meso and macro levels rather than just the micro (Fig. 4.1). This ecosystem approach will recommend that higher education educators work collaboratively across society with diverse stakeholders so as to prepare students for the future of work.

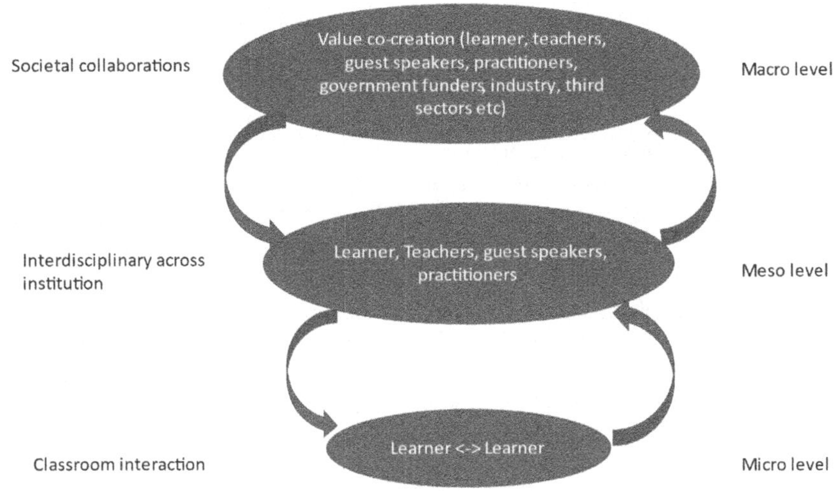

Fig. 4.1 Proposed service-dominant logic in the higher education ecosystem (Adapted from Barile et al., 2016)

Service-Dominant (S-D) Logic and Its Relevance

Vargo and Lusch (2004) first developed service-dominant (S-D) logic as an alternative explanation to traditional logics of exchange, with an emphasis on humans applying their competencies to benefit others and benefitting reciprocally from others' application of their competence. S-D logic fundamentally places systemic value co-creation with its various actors at its core. Over time, Vargo and Lusch have developed and expanded the concept of service-dominant logic into five axioms and the concept of a service ecosystem (Lusch & Vargo, 2014).

Using the S-D logic concepts relating to the value creation potential in the arrangements of actors and institutions (Vargo & Lusch, 2016), it seems both beneficial from a knowledge development perspective, and a foundational approach when examining the relationships and value creation in the higher education service ecosystem. Whilst there has been an evolution of the S-D logic concepts and application with an acknowledgement of the complex systems approach, we felt that it would be appropriate to explore the value creation ecosystem for higher education specifically regarding the anticipated influence of Industry 4.0 on graduate skills development which has been accelerated due to the Covid-19 pandemic. Figure 4.1 outlines the application of service-dominant logic in the higher education ecosystem. This diagram depicts how we view the various levels of learning taking place and the importance of creating an inclusive learning environment where ultimately value co-creation takes place through collaborative actions.

To investigate further poses the questions, what would a new world of the HEI services ecosystem look like? Who are these actors and how will the value co-creation be manifested? It is important to recognise the social and political context of the service ecosystem that supports the exchange and interaction of resources between the actors which can act to increase the value derived (Akaka & Chandler, 2018). In relation to the provision of higher education, there is a multiplicity of actors engaged in the system of provision from learners, employers, statutory and professional bodies, even parents or tuition fee-payers. This systemic complexity demands an increased investment in the mapping of this ecosystem in order for HEIs to be able to increase the value-in-use of learning for stakeholders and to leverage the opportunity presented by transdisciplinary alignment (Budwig & Alexander, 2020; Piaget, 1972).

S-D Logic and Higher Education

Within the UK higher education environment, there has been emerging recognition of the importance of value creation interdependency amongst stakeholders. This has resulted in the diminishing of conventional boundaries between education disciplines, providers and employers. In the UK in particular, there have been positive developments in interdisciplinary degree programmes across departments, schools and even at institutional level (e.g. University of Aberdeen's interdisciplinary centres); cross-delivery of programmes by different education providers (e.g. Central Saint Martins-Birkbeck MBA programme) and co-creation of educational programmes by employers with educational institutions (e.g. apprenticeship degree partnerships). It can be said therefore, that the relationship between the learner and the learning organisation, as an adaptive system, in higher education can be complex and evolving, as the learner or student role develops from a passive recipient of 'goods or services' (i.e. learning) to the active customer or client role (Bailey, 2000) with the associated additional value expectations.

UK Higher Education Context

UK Higher Education Institutions (HEIs) have gone through over a decade of tumultuous changes including changes in funding models (Browne, 2010), research funding cuts particularly as a result of exiting the European Union in 2016, pressures to enhance contributions to the 'knowledge society' and heightened vocalisation of expectations regarding teaching (Gunn & Fisk, 2013). As we write this chapter, the global pandemic is still ongoing, with new variants emerging and continuous and spontaneous lockdowns occurring and ever-changing rules and introductions of hybrid, blended teaching and learning opportunities, some more successful than others. The most recent review of UK HEIs by Lord Augar in 2021 (Pickford, 2021) indicated that the direction the UK government was taking is around lifelong learning, recommending that the government "should introduce a single lifelong learning loan allowance for tuition loans at Levels 4, 5, and 6, available for adults aged 18 or over... equivalent to four years' full time undergraduate degree funding".

According to the World Economic Forum's Future of Jobs Report (2020), half of all employees will need reskilling by 2025 as a result of an increase in the adoption of technology. The ongoing pandemic has pointed to this and if anything, revealed the technological gap in households as workers and their children turned to home-schooling whilst educational establishments moved online in an effort to stem the spread of the pandemic. In the same report, skills such as critical thinking and problem-solving were on the top of the list of skills employers thought will need to be in employees' skill sets by 2025. What does this mean then, for learning and teaching content and pedagogy? Importantly, how do we work internally within universities (students and faculty) and externally with our partners and collaborators (employers, government, other funders) to develop a collaborative education, work and skills ecosystem?

Innovative Practices in UK Higher Education Institutions

In the next section of our chapter, we will share some examples of innovative practices that we have encountered that incorporate elements of this future skill building, in a collaborative manner that involves various stakeholders. We suggest that success in building such ecosystems is where the triple helix of academia, employers and government work proactively and collaboratively to develop such future skill sets (Etzkowitz & Leydesdorff, 1995). What all these examples demonstrate are active attempts to engage with teaching and learning activities which contribute to graduate attributes and capabilities that think tanks claim are the top skills required for the near future.

With reference to Raworth's (2018) and Mazzucato's (2019) seminal pieces, we suggest that Higher Education Institutions need to employ the pedagogy of value co-creation—the urgent need for diverse stakeholder groups within greater society to rethink the concept of value (not just a narrow economic output).

Complex Problems and Problem-Solving as a Centre Piece

The Central Saint Martins-Birkbeck (CSM-BBK) Masters of Business Administration was established in 2016 and launched in 2017. It is currently in its 5th cohort. It was established as the first MBA in the world between a design school and a business school to address global and societal challenges. The programme was developed in response to an increasing need for creative and critical approaches to the leadership and management of enterprises in order to facilitate positive organisational change. With reference to Fig. 4.1, we suggest that the case study of the CSM-BBK MBA demonstrates the exploitation of the potential provided at the meso and macro levels, where interaction goes beyond the classroom interaction and involves interdisciplinary subject areas and the wider community and society.

Most conventional MBAs offer ineffective responses to uncertainty, technological revolution and globalisation, with institutions finding it difficult to innovate sufficiently if constrained by disciplinary structures (Roberts, 2015). Fahim et al. (2021) specifically note the decline in MBA enrolments in major schools around the world and contend that this derives in part from the difficulty in identifying the value to the student from a costly business programme (Wilson & Thomas, 2012). CSM and Birkbeck recognised the need to move away from a discipline-centred pedagogy towards a student-centred approach based on a social constructivist perspective. Drawing together academics from a range of arts and business disciplines, a practice-focused, values-based programme emphasising the co-production of knowledge was developed (Walsh & Powell, 2020).

From the inception of the CSM-BBK partnership, the programme designers were committed to developing an MBA programme which was radically different, using a social constructivist pedagogy, drawing on theories of situated experience which give primacy to the dynamics of everyday existence, improvisation, coordination and interactional choreography. They mostly address the interactive relations of people with their environment (Wenger, 1998). The commitment to the social creation of knowledge is fully integrated into programme design activities, the structure of which echoes the 'sprints' which the students would experience. In developing the programme, academic and practitioner-oriented ('pracademics') colleagues from both institutions met for extended blocks of time over several months with the intention of producing an output (a module outline) at the end of each interaction. The concentration of contact (as opposed to a standard meeting format) enabled a deep exchange of disciplinary perspectives, so that there could be integration rather than a relatively superficial 'joining up' of approaches. This intertwining of the distinct pedagogic approaches of business and the arts allowed the programme to draw on the respective strengths of the disciplines involved, enabling students to enhance their professional practice, not simply learn more about practice generally.

The approach taken challenges the widespread assumption in higher education that practice involves the application of theory that has been previously acquired, and the distinction frequently made between learning in the academy and learning in, and from, the workplace. Eraut (2004) points out that professional and managerial performance involves the simultaneous use of several types of knowledge, which have to be learned holistically.

The creation of a problem-centred, systems-thinking, flipped classroom approach enabled a move away from the traditional information transmission model, with students as passive learners, to support an active learning environment where students apply concepts and tools in context, drawing on their professional and practice knowledge. The delivery model emphasises active and experiential learning, which is a form of embodied learning (i.e. it involves the whole person) and, therefore, provides a contrast to the rational cognitive model of learning which is dominant in most management approaches to higher education. They combine previously delivered 'taught' material with opportunities to put into practice the knowledge and skills that have been acquired. There is a strong emphasis on group production of knowledge, and the first extended module introduces students to agile development, in which requirements and solutions evolve through collaboration between self-organising, cross-functional teams. The emphasis on group experience alerts students to the extent to which knowledge in the workplace is socially produced through people working together. It is clear that post-experience, post-graduate individuals, such as MBA students, benefit from reading/learning in their own time (as they are in work) and attending workshops face-to-face where they actively engage with their classmates and the tutors, guest speakers and senior leaders of businesses.

The programme uses an amalgam of the creativity processes from the arts and design school tradition (for example, action research, conceptual prototyping and iteration) and analytical skills from business and management disciplines. Adopting a 'team teaching' approach using members from each institution, the synergy created through the development of this community of practice between the two institutions is intended to deliver significant organisational learning as the programme matures. The hybrid teaching and learning approaches enabled students to develop a critical stance to their own practice and to the context in which they operate. The programme focuses on alternatives to the dominant value systems, foregrounding social, ethical and environmental values and introducing students to a much wider awareness of markets and contexts. In contrast to encouraging students to view the world of practice as one of the neat problems with ready solutions, the programme supports them in working with 'wicked problems'—"a class of social systems problems with a fundamental indeterminacy without a single solution, and where much creativity is needed to find solutions" (Johansson-Sköldberg et al., 2013). These wicked problems are worked through together with societal collaborators and organisations.

Engagement with arts and business traditions also gives students a distinctive perspective, in that exposure to the methods of different disciplines leads them to compare not just methods, but the epistemologies underlying the methods, with the result that diversity frames research orientations as particular socially constructed perspectives—not sacrosanct pathways to the truth, so that all methods are subject to questioning and analysis (Kincheloe, 2001). The innovative pedagogy is reflected in the assessment, and the assessment criteria used. In order to support the ethos of the MBA, with its focus on knowledge building and learning, much of the work undertaken is formative. Formative assessments are wide-ranging and include group and individual presentations, elevator pitches, case study analyses, reports, videos and rich picturing. Formative assessments allow students to learn and develop their understanding and practice without having a direct impact on their final unit grade. Unusually for an MBA, the programme includes just four summative assessments.

As we can see, developing problem-solving strategies is clearly embedded through the iterative manner of the learning and foundational building of knowledge. In utilising the UN Sustainable Development Goals as a guide for the various complex societal challenges, each unit or cohort of students approaches these global challenges in an attempt to get closer to providing workable solutions (grand or otherwise) with relevant organisations.

Experiential Learning, Employability and Work-Based Learning

Another example of a co-created programme would be that of the degree apprenticeship; programmes offered by some universities in the UK, which are developed by employers, universities and professional bodies working in partnership (UCAS, n.d.). Students typically are employed throughout the programme and split their time between the university and their employers. In fact, modern apprenticeships can be said to be the most integrated form of work-based learning (QAA, 2018) as they have been established to meet an identified workplace skills gap.

A central tenet of the degree apprenticeships is that of a tripartite arrangement where the learner (apprentice), learning provider (HEI) and employer collaborate to create knowledge, develop skills and deliver value in the form of enhanced workplace behaviours. We would contend that this approach is supporting the development of an ecosystem of resource integration where multiple actors are engaged in dynamic collaboration and co-operation as a progression away from the traditional didactic student–educator relationships isolated from employer needs.

In S-D logic, Axiom 3 (Vargo & Lusch, 2017) identifies that the actors concerned with social and economic aspects of the service are collectively driven by the need for knowledge and skills (Akaka & Chandler, 2018) and will seek access to these rather than just ownership. Therefore, harnessing the actors to co-create value in higher education through the construction of an apprenticeship standard and associated framework further supports this collaborative learning ecosystem.

In the UK, the government has sought to address the identified gaps in business skills in post-16 education including at the HE level through the introduction of the degree apprenticeships in 2015–2016 developed in partnership between HEI, Employers and professional bodies (Hubble et al., 2019). Each degree apprenticeship is developed from the relevant discipline's apprenticeship standard, and the funding for a programme of study is sourced through the national apprenticeship levy (Hubble et al., 2019). The core driver in the development of apprenticeship standards was the requirement for employer-led design aligned with industrial strategy (Universities UK, 2019). The establishment of these 'trailblazer' groups to design the occupational standards in consultation with training providers (e.g. further education colleges, Higher Education Institutions, etc.) and industry professional bodies has enabled the provision of around 766 UK occupational standards across levels 2–7. Each occupational standard uses a framework of knowledge, skills and behaviours as the basis for identifying outcomes from an apprenticeship and therefore subsequent value created for the time and money invested by those in the tripartite arrangements.

Taught modules within degree apprenticeships modules delivered at the Christ Church Business School are structured using the relevant apprenticeships occupational standard (Institute for Apprenticeships & Technical Education, n.d.). As there is still a requirement for teaching quality to be maintained for the degree element of the apprenticeships, mapping the required knowledge, skills and behaviours to validated module learning outcomes is a detailed and rigorous process involving a tripartite arrangement between the learner, the workplace and the learning provider (HEI). The resultant course learning design is grounded in experiential learning (Kolb, 1984) utilising the associated techniques of reflection on practice, active learning and feedback on the learning experience, combined with active learning methods such as role-play, group work and case study evaluation.

The Chartered Manager Degree Apprenticeship (CMDA) standard is predicated on a learning design that encompasses on-the-job and off-the-job learning where apprentices develop a learning community that is both within the workplace and within the classroom. The situated learning aspect, referred to as learning through doing (Lave & Wenger, 1991), of this pedagogy provides the apprentice with the opportunity to experience a workplace challenge, and to utilise this example for reflection and subsequently develop knowledge and skills relative to the experience with a change in relevant workplace practice. The CMDA course design is underpinned by an assessment structure that in the first instance assesses knowledge on a management topic, then requires the apprentice to apply this knowledge to a current challenge within their workplace. This methodology culminates in the final professional project, an extended report, developed by the apprentice in conjunction with their workplace mentors and managers. The iterative development of knowledge and skills continues as the apprentice, in order to complete the apprenticeship standard after the traditional degree qualification, is required to present an independent assessment panel with detail of the professional project as a core component of the end-point assessment (EPA). The EPA is an independent (from the learning provider) assessment of the extent to which the apprentice has mastered the knowledge, skills and behaviours required, the culmination of the tripartite collaboration.

Interdisciplinary Centres of Excellence

In preparing for the skills of the future (complex problem-solving being one of them), interdisciplinary centres in forward-looking institutions have been set up as part of their long-term strategy. We suggest that these interdisciplinary centres of teaching and research excellence are fine examples again, of how service-dominant logic works in enabling teaching, research and knowledge exchange to happen at all three levels (micro, meso and macro) of learning. The University of Aberdeen's recent announcement of the setting up of five interdisciplinary Centres—"hubs designed to encourage and enable academic teams to work across scientific boundaries to provide solutions to complex problems facing the key challenges of our time, through innovative research and education" (University of Aberdeen, n.d.) is one such example of educational leadership. The effective management of this subsequent ecosystem of learning could ensure transformative learning experiences for students, deliver relevant skills improvement for employers and ground the positive contribution HE makes to regional, national and international economies, ultimately contributing to what Piaget (1972) called "the self-renewal" of society. In their strategy document titled Aberdeen 2040, amongst the various purposes outlined (e.g. education and research, inclusivity and sustainability), one which resonated with us was that of the identification of their own interdisciplinary challenges, which they have committed to addressing. As they put it, "these urgent and wide-ranging problems require complex solutions, which draw on both theoretical and applied knowledge" (University of Aberdeen, n.d.). These are listed as energy transition, health, nutrition and wellbeing, data and artificial intelligence, environment and biodiversity and social inclusion and cultural diversity.

Within each of the recently established centres, it is clear that the emphasis is on being inclusive of a diverse range of subject specialisms and embracing both traditional and alternative methods of understanding, learning and experience. For example, within the interdisciplinary Centre of Social Inclusion and Cultural Diversity, they talk about employing diverse academic and applied research and knowledge to address complex challenges of the present. They feature medieval archives, extensive literature collections, musical collections and war memorials, the use of digital technology to bring to life historical pasts, understanding immigrant children and families and understanding politics through diverse perspectives. Much of the research does not appear to sit 'nicely' within a siloed, narrow definition of a traditional single-subject area. Rather, much of the impact has come about through collaborations between two or more subject specialisms, resulting in bridging perspectives and gaps and the opening up of important conversations.

Conclusion

In this chapter, we have established that the UK higher education sector continues to undergo tremendous changes and as with complex challenges, many stakeholders at different levels are involved in the value generation and co-creation journey. Complexity in global challenges, as identified by the UN Sustainable Development Goals are difficult challenges to solve or resolve. They require a combined and coordinated global effort from multiple partnerships and approaches in order to even attempt addressing these challenges. Within the seventeen UN SDGs that were created and adopted by all UN Member States in 2015, it was recognised in the shared blueprint and united call for action, that "ending poverty and other deprivations must go hand-in-hand with strategies that improve health and education, reduce inequality and spur economic growth—all while tackling climate change and working to preserve our oceans and forests" (United Nations Department of Economic and Social Affairs, n.d.).

Recent publications such as Raworth's (2018) *Doughnut Economics* and Mazzucato's (2019) *The Value of Everything* reflect the slow transformation and understanding that society and its people need to rethink the concept of value. Specifically, such modern takes on new economic structures encourage the understanding that current modern capitalistic value-extracting concepts need to be debunked in favour of a more sustainable co-created, value-creating system that works for everyone.

We have shared a few exemplars of how, in adopting S-D logic axioms, it is entirely possible to deconstruct and re-construct teaching curriculum and pedagogies to ensure that students in Higher Education Institutions can benefit from programmes that have had constructive inputs from multiple experienced and knowledgeable stakeholders. Complexity in global challenges demands a systemic, less siloed approach to create shared value from the stakeholder ecosystem. Through collaboration with the actors, we can start to rethink fresh approaches to the change leadership role that HEIs can perform.

References

Akaka, M., & Chandler, J. (2018). Reframing exchange: A service-ecosystems perspective. In S. L. Vargo & R. F. Lusch (Eds.), *The Sage Handbook of service-dominant logic* (pp. 135–148). SAGE Publications.

Bailey, J. J. (2000). Students as clients in a professional/client relationship. *Journal of Management Education, 24*(3), 353–365.

Barile, S., Lusch, R., Reynoso, J., Saviano, M., & Spohrer, J. (2016). Systems, networks, and ecosystems in service research. *Journal of Service Management, 27*(4), 652–674.

Browne, J. (2010, October 12). *Securing a sustainable future for higher education.* https://www.gov.uk/government/uploads/system/uploads/attachment_data/file/422565/bis-10-1208-securing-sustainable-higher-education-browne-report.pdf

Budwig, N., & Alexander, A. J. (2020). A transdisciplinary approach to student learning and development in university settings. *Frontiers in Psychology, 11*, 576250.

Eraut, M. (2004). Non-formal learning and tacit knowledge in professional work. *British Journal of Educational Psychology, 70*, 113–136.

Etzkowitz, H., & Leydesdorff, L. (1995). The triple helix – University-Industry-Government relations: A laboratory for knowledge based economic development. *EASST Review, 14*(1), 14–19.

Fahim, A., Addae, B. A., Ofosu-Adarkwa, J., Tan, Q., & Bhatti, U. A. (2021). Industry 4.0 and higher education: An evaluation of barriers affecting master's in business administration enrolments using a grey incidence analysis. *IEEE Access, 9*, 76991–77008.

Fuller, A., & Unwin, L. (1998). Reconceptualising apprenticeship: Exploring the relationship between work and learning. *Journal of Vocational Education & Training, 50*(2), 153–173.

Gunn, V., & Fisk, A. (2013). *Considering teaching excellence in higher education: 2007–2013: A literature review since the CHERI report 2007. Project report.* Higher Education Academy. http://eprints.gla.ac.uk/87987/1/87987.pdf

Hubble, S., Bolton, P., & Foster, D. (2019). *The forthcoming review of post-18 education and funding*. House of Commons Library. https://researchbriefings.parliament.uk/ResearchBriefing/Summary/CBP-8490

Institute for Apprenticeships & Technical Education. (n.d.). *Occupational standards*. https://www.instituteforapprenticeships.org/occupational-standards/

Johansson-Sköldberg, U., Woodilla, J., & Çetinkaya, M. (2013). Design thinking: Past, present and possible futures. *Creativity and Innovation Management, 22*(2), 121–146.

Kincheloe, J. L. (2001). Describing the Bricolage: Conceptualizing a New Rigor in qualitative research. *Qualitative Inquiry, 7*(6), 679–692.

Kolb, D. A. (1984). *Experiential learning: Experience as a source of learning and development*. Prentice-Hall.

Lave, J., & Wenger, E. (1991). *Situated learning: Legitimate peripheral participation*. Cambridge University Press.

Lusch, R. F., & Vargo, S. L. (2006). Service-dominant logic: Reactions, reflections and refinements. *Marketing Theory, 6*(3), 281–288.

Lusch, R. F., & Vargo, S. L. (2014). *Service-dominant logic: Premises, perspectives*. Cambridge University Press.

Mazzucato, M. (2019). *The value of everything: Making and taking in the global economy*. Penguin.

Piaget, J. (1972). Intellectual evolution from adolescence to adulthood. *Human Development, 15*(1), 1–12.

Pickford, R. (2021, June 2). *Mapping the policy influence of Augar: How many of the 53 recommendations have been addressed by the Government?* Higher

Education Policy Institute. https://www.hepi.ac.uk/2021/06/02/mapping-the-policy-influence-of-augar-how-many-of-the-53-recommendations-have-been-addressed-by-the-government/

QAA. (2018). *UK quality code for higher education, advice and guidance, work-based learning (The Quality Assurance Agency for Higher Education)*. UK Standing Committee for Quality Assessment. https://www.qaa.ac.uk/docs/qaa/quality-code/advice-and-guidance-work-based-learning.pdf?sfvrsn=f625c181_2

Raworth, K. (2018). *Doughnut economics: Seven ways to think like a 21st century economist*. Cornerstone.

Roberts, P. (2015). Higher education curriculum orientations and the implications for institutional curriculum change. *Teaching in Higher Education, 20*(5), 542–555.

Senge, P. (1990). *The fifth discipline, the art and practice of the learning organization*. Doubleday/Currency.

UCAS. (n.d.). *Degree apprenticeships*. UCAS. Retrieved March 16, 2022, from https://www.ucas.com/apprenticeships/what-you-need-know-about-apprenticeships/degree-apprenticeships-0

United Nations Department of Economic and Social Affairs. (n.d.). *Do you know all 17 SDGs?* https://sdgs.un.org/goals

University of Aberdeen. (n.d.). *Interdisciplinary challenges*. University of Aberdeen. Retrieved March 16, 2022, from https://www.abdn.ac.uk/2040/interdisciplinary-challenges/index.php

Universities UK. (2019). *The future of degree apprenticeships*. Universities UK. Retrieved March 16, 2022, from https://www.universitiesuk.ac.uk/sites/default/files/field/downloads/2021-07/future-degree-apprenticeships.pdf

Vargo, S. L., & Lusch, R. F. (2004). Evolving to a new dominant logic for marketing. *Journal of Marketing, 68*(1), 1–17.

Vargo, S. L., & Lusch, R. F. (2016). Institutions and axioms: An extension and update of service-dominant logic. *Journal of the Academy of Marketing Science, 44*, 5–23.

Vargo, S. L., & Lusch, R. F. (2017). Service-dominant logic 2025. *International Journal of Research in Marketing, 34*(1), 46–67.

Walsh, A., & Powell, P. (2020). Re-imagining the MBA: An arts-based approach. *Higher Education Pedagogies, 5*(1), 148–164.

Wenger, E. (1998). *Communities of practice: Learning*. Cambridge University Press.

Wilson, D. C., & Thomas, H. (2012). The legitimacy of the business of business schools: What's the future? *Journal of Management Development, 31*(4), 368–376.

World Economic Forum. (2020). *Future of jobs report*. https://www.weforum.org/reports/the-future-of-jobs-report-2020

CHAPTER 5

Embracing a Pedagogy of Ambiguity in Higher Education

Veronika Kelly

Introduction

The World Economic Forum's *The Future of Jobs Report 2020* points to a "highly uncertain outlook for the labour market" exacerbated by the Covid-19 pandemic and associated global recession. According to this report, key skills needed in 2025 include critical thinking, analysis, and problem-solving coupled with a capability to thrive in uncertain and complex work situations—all skills in the realm of human ingenuity. Additionally, the share of tasks performed by people, as opposed to machines, expected in 2025 is also highest in communicating and interacting, reasoning and decision-making, coordinating, developing, managing, and advising. While criticality, communication, and problem-solving skills are central to student learning outcomes in universities, the forecast for the

V. Kelly (✉)
University of South Australia, Adelaide, South Australia, Australia
e-mail: veronika.kelly@unisa.edu.au

M. J. Lehtonen et al. (eds.), *Design Education Across Disciplines*,
https://doi.org/10.1007/978-3-031-23152-0_5

future of work beyond known routines and limits has additional implications for higher education and how it functions, and in turn, what university learning is good for.

Higher education in Australia is a complex and changeable environment. Shifting student demographics, the expansion of digital learning, diversification of student markets, changes in federal government higher education policies and priorities, and reductions in funding all impact on university teaching and learning. The impacts of the global pandemic have been severe and brought widespread challenges to higher education. Many institutions had to act quickly to adapt on-campus practice-based teaching to alternative approaches in situations that were continually changing, revealing varying degrees of preparedness. For design education, where the certainty of the studio as a physical site of learning and social interaction retains its power for many design educators, shifting modes of engagement to virtual, distributed formats has had implications for the ways in which educators and students engage with each other (Marshalsey & Sclater, 2020). In architecture education, this shift has been referred to as akin to the grief process, one of transformation "from denial to acceptance" (Brown, 2020). The challenges around online learning, particularly in areas such as studio and practice-based learning have also ignited opportunities for reflection (Jones & Lotz, 2021). Many design educators had to re-evaluate "traditional" studio learning and their own roles in order to teach design virtually, affording the potential for reflexivity in teaching and learning at a global scale. Virtual studio teaching combined with a higher order focus on inclusivity and engagement in shared learning environments further revealed the diverse voices of students that can be hidden in physical studio settings (Gray, 2021). These experiences are enabling design educators to critically question the values and embedded knowledge implicit in their own teaching practices towards new learning conventions (Fleischmann, 2021; Jones & Lotz, 2021; Kelly et al., 2021; Lehtonen et al., 2021).

The past two years have also afforded a potential to recognise the appeal of uncertainty as a vehicle for opportunities (Nowotny, 2016), and of ambiguity as a function of and a resource for learning. For higher education, this period can be characterised as continually working through ambiguity, requiring rapid, iterative responses to the dynamic and evolving impacts of the pandemic. Designers and design educators are familiar with working with uncertainty and complexity; design is after all concerned with the prospective and with possibility, when there is no "perfect plan or design" (Jones & Lotz, 2021, p. 5). In discussing co-creating virtual design studios through a lens of play Lehtonen et al. (2021) also highlight how ambiguity, combined with the pressure of an immediate shift to online learning, enabled new ideas to emerge. The pandemic therefore brought not only a focus on the utility of uncertainty and ambiguity in the design process and for design education; it highlighted a need for *all* educators to anticipate (and even embrace) ambiguity as they interpreted their teaching practices differently to continue supporting student learning. These global challenges highlight the relevance of design's signature pedagogies for teaching and learning in higher education. Ambiguity and uncertainty are conceived as central concepts in design education and particularly in the context of practice/studio-based learning. Ambiguity has been positioned as a productive force that supports learning, indicative of a capacity to formulate different interpretations of one thing rather than being fixed on one answer (Orr & Shreeve, 2018). Although a familiar concept in teaching design and design practice, ambiguity has significance for teaching and learning in higher education more broadly, where supporting students to develop self-assuredness in relation to uncertainty will better prepare graduates for uncertain employment contexts (Orr & Shreeve, 2018). In this regard, the way that educators approach facilitating ambiguity in their local setting is integral to providing this support to students.

Austerlitz et al. (2008) used the term *pedagogy of ambiguity* in referring to art and design education, and it is proposed here as a way of assisting all higher education students to prepare for the uncertain world of work. In this chapter I draw on empirical work from different disciplines and observations from design and design's signature pedagogies, to reflect on the discourse of ambiguity across different fields. The chapter further illustrates that the value of ambiguity in design education has relevance for the changing nature of work and careers, and through embracing a pedagogy of ambiguity, for developing students' and educators' transformative potential more broadly. In doing so, this chapter does not seek to provide a single framework or "toolbox" for teaching ambiguity, but to generate a series of provocations for educators to explore in their own teaching practices. Embracing ambiguity as a productive and intentional activity is proposed as a way for educators to expand knowledge towards new possibilities for teaching and learning, while critically questioning their own teaching practices and approaches to student learning.

This chapter begins with a brief discussion of higher education challenges today in setting the scene for the value of ambiguity to teaching and learning; then traces the ways in which ambiguity has been discussed in fields such as psychology and organisational management, and then compares this to the conceptualisation of ambiguity in design and design education, and in relation to design's signature pedagogies. The chapter conceives of ambiguity as a discourse that is culturally and socially constructed in its context of use and concludes with a series of prompts for a pedagogy of ambiguity to support the transformation of teaching and learning in higher education today.

Higher Education, Uncertainty, and Ambiguity

Higher education institutions are increasingly challenged by complexity, change, and ambiguity; their futures are uncertain, and they are subject to external scrutiny, growing competition, and financial challenges (Buller, 2014; Dunican et al., 2019; Marshall et al., 2022). External structural imperatives and drivers include government priorities and performance measures based on teaching quality, student satisfaction, graduate employment outcomes, and employer satisfaction (Austerlitz et al., 2008; Dunican et al., 2019). Internally, there is increased regulation, risk management and controls, and the establishment of performance indicators, where policies and their formulation "are a crucial site of institutionalised forms of coping with uncertainty" (Nowotny, 2016, p. 121). The pursuit of clarity is paramount, and the materialisation of risk as an unknown in organisations has seen it change "into something known by converting danger into a risk that could be calculated and hence contained" (p. 65). Yet dealing with uncertainty is messy, and impermanent, and strategies can range from those that are routinised for everyday decisions, to being "ad hoc and haphazard" (p. 120).

Within this shifting and uncertain context, a complex and ranging combination of teaching, research, service, and external outreach is undertaken. Adding to this complexity is the ongoing tension between the social contribution of education and the marketisation of higher education. A neoliberal view of higher education positions universities "as a financial investment for the student and a vehicle for serving the needs and demands of the business community" (Mintz, 2021, p. 82). Yet higher education is distinct from business; the creation, dissemination, translation, and application of "knowledge through multiple means for many different audiences and purposes is exciting and significant work, but it is not easy, and outcomes are often difficult to observe and assess" (Bolman & Gallos, 2021, p. 8). The very mission of discovering and disseminating knowledge or creating educated and responsible citizens and professionals is by nature ambiguous. For incoming academic leaders tasked with interpreting their environment, their "success at deciphering ambiguity is essential in order to make decisions on what to change" (Dunican et al., 2019, p. 41). In higher education ambiguity is both a risk to be managed and a positive force to be harnessed for creating opportunities and providing "wiggle room" for skilled leaders (Bolman & Gallos, 2021, p. 128). This "wiggle room" comes back to the ability to consider and generate different interpretations of one thing or situation.

Beyond higher education, O'Connor et al. (2018) flag that in addition to organisations dealing with increased competition, technological advances, and globalisation, their employees are also facing growing uncertainty as their roles are expanding alongside internal forces, such that employees "are regularly exposed to unclear and ambiguous situations" (p. 2). This increased ambiguity signifies employees "who can *tolerate* or even *embrace* ambiguity at work will be well placed to perform well in contemporary organisations" (p. 2). There are implications here for higher education in recognising the value of teaching ambiguity as "today's graduates need to understand multiple perspectives holding competing views simultaneously" (Orr & Shreeve, 2018, p. 63). It also has implications for educators in higher education more broadly and increases a learner's ability to embrace ambiguity as a fact of life and force for change rather than something negative to be avoided or overcome (Visser & Visser, 2004).

Before turning to the ways in which knowledge of ambiguity has been produced in the literature from different fields, it is useful to consider the concept of discourse in the production and (re)production of knowledge. Discourse is commonly conceived as socially situated language use. However, language and its use can also be considered as "a form of social practice, rather than a purely individual activity or a reflex of situational variables" (Fairclough, 1993, p. 63). In this regard, the use of language—consisting of all types of accounts, or "texts"—is conceived as a social construction whether spoken, written, or visual; it is discourse. Discourse, when conceived as a social practice, is simultaneously a method of understanding and signifying the world, a mode of acting upon the world and other persons, and also a means of transforming these operations (Fairclough, 1993, 2002). In these terms, knowledge of ambiguity is produced in its culture of use, and it both shapes and is shaped by culture and the wider discourses from which it draws. Therefore, knowledge of ambiguity can be understood as produced by people in a particular culture (e.g., a learning environment, an organisation, a design studio) and underpins how knowledge is expressed, constituted, and legitimised, or in other words, accepted as knowledge. And because discourse is a means of production, reproduction, and regulation of knowledge, there is also the potential for transformation. Interpreting ambiguity as a discourse in its specific culture of use therefore presents an opportunity to transform how ambiguity is understood and approached.

Interpreting Ambiguity

The terms ambiguity and uncertainty have blurred boundaries and are frequently used interchangeably. In the *OED*, uncertainty as a noun is a *quality* ("the quality of being indeterminate"), and a *condition* (the state of "not being perfectly clear", "doubtfulness or vagueness"). Ambiguity as a noun, defined in reference to language, is a *quality* ("quality of having different possible meanings", "being interpreted in more than one way"). While uncertainty can refer to indeterminacy, ambiguity is different in that it pertains to openness to alternative ideas or multiple interpretations (Eco, 1976; Orr & Shreeve, 2018; Suzawa, 2013). In these terms, ambiguity points to the interpretative relationship between people and things or situations (Gaver et al., 2003), and to culture and context. This is the lens that I bring to ambiguity and its relevance to teaching and learning in higher education.

The concept of ambiguity has been widely researched in fields such as psychology, organisational management and leadership, education, and design education. In psychology, discussions centre on ambiguity tolerance as a personality trait or variable, and in design education ambiguity is discussed as a threshold concept, a necessary ingredient of the design process and developing learners' creativity. These different ways that ambiguity is discussed enable individuals to adopt different positions in relation to ambiguity, and construct knowledge of ambiguity through discourse. A review of literature identifying and key themes from these fields, and through a lens of ambiguity as discourse, is useful for understanding how ambiguity in design education contributes to teaching and learning more broadly.

Tolerating Ambiguity

A tolerance of ambiguity as a subject of research in psychology emerged in the mid-twentieth century and has extended to organisational management, leadership, and education, including drawing on research participants from higher education (Budner, 1962; Buller, 2014; Dunican et al., 2019; McLain, 2009; Norton, 1975). The concept of an intolerance of ambiguity originated as an "emotional and perceptual personality variable" (Frenkel-Brunswick, 1949) and later as "the tendency to perceive (i.e., interpret) ambiguous situations as sources of threat" and the tolerance of ambiguity as "the tendency to perceive ambiguous situations as desirable" (Budner, 1962, p. 29). Much of this early research is based on psychometric testing to measure tolerance of ambiguity as a personality variable. In these empirical studies, the intolerance/tolerance of ambiguity refers to the evaluation made by a respondent, rather than studying the explicit behaviours performed in response to ambiguity. A vast body of research has sought to extend the measure and application of ambiguity tolerance, increasingly shifting from measuring a tolerance of ambiguity as a personality trait to a more context-specific construct (Furnham & Marks, 2013).

At the same time, the concept of a "tolerance of ambiguity" has continued to have primacy, including in higher education and in relation to leadership (Dunican et al., 2019; Herman et al., 2010; McLain, 2009; Norton, 1975; O'Connor et al., 2018). For example, the ability to interpret ambiguity is proposed as a central skill for academic leaders making decisions on what to change, with links made between lower ambiguity tolerance—bigger resistance to change, and greater tolerance of ambiguity—more positive attitude to change (Dunican et al., 2019). However, the authors recognise that knowledge of contextual and cultural influences of attitudes towards resistance is a limitation of this study. (O'Connor et al., 2018) look to the broader workplace in studying the relation between a person's tolerance of ambiguity and adaptive behaviours at work (p. 16). The authors propose that ambiguity tolerance is a predictor of creativity and an important leadership quality, beneficial in the context of growing ambiguity and uncertainty at work. Ambiguity tolerance also appears in personality tests such as the Hogan Personality Inventory, used by organisations in staff recruitment and professional and leadership development.

Fairclough (1993) points to the importance of historical change in relation to discourse and "how different discourses combine under particular social conditions to produce a new, complex discourse" (p. 4). For example, the discourse of ambiguity tolerance as a personality trait to be measured problematises the individual at the centre. A person who is more tolerant of ambiguity is perceived as more motivated to learn and engage, or less tolerant of ambiguity is more worried, anxious, or change resistant. The discourse of ambiguity tolerance also emerges in organisations as a predictor of the ways that a person behaves or adapts to situations as an indicator of resistance to change, work performance, creativity, and leadership skill (Dunican et al., 2019; Herman et al., 2010; O'Connor et al., 2018). These are examples of the ways in which discourse shapes social constructions of ambiguity tolerance as a desirable personality trait for contemporary work, thereby creating a new discourse of ambiguity tolerance and legitimising it as knowledge. The discourse of a tolerance of ambiguity, and of uncertainty, has also appeared in design education in relation to creativity and to threshold concepts (Mahmoud et al., 2020; Osmond & Turner, 2010).

Ambiguity in Design Education

The design studio is widely acknowledged globally as a site of learning through which the material, critical, project-based, and dialogic ways of teaching and learning design are enacted. As design's signature pedagogies, these ways of knowing and doing support learners to "become" designers, enculturated into a design profession and community of practice (Shreeve, 2012, 2015). Shulman (2005a) describes signature pedagogies in higher education as pervasive, habitual, and routinised, and as such they "permit students to spend far less time figuring out rules of engagement, which enables them to focus on increasingly complex subject matter" (p. 22). Signature pedagogies comprise three structural dimensions: "surface" (tangible and operational teaching and learning activities); "deep" (assumptions about the best way to convey knowledge/expertise), and "implicit" (a set of beliefs regarding professional values, attitudes, and character) (Shulman, 2005b, pp. 54–55). As such, these routinised behaviours, practices, and structural dimensions contribute to the constitution of design knowledge and cultures of practice, and support student learning by establishing modes of interacting and operating in the learning environment.

Ambiguity and uncertainty permeate these signature pedagogies as threshold concepts. For example, working with uncertainty is described as a condition essential to the design process (Jones, 2020); it is a means of working through what is not "known" to some way of dealing with the unknown, and where a tolerance of uncertainty in design is deemed transformative for learners (Osmond, 2009). Similarly, pedagogies of uncertainty are described as being "where students are learning how to act under conditions where knowledge is limited yet actions must be taken" (Shulman, 2005a). A threshold concept implies therefore a transition from one point to another—which is really at the heart of what all learning is. Meyer and Land (2005) introduced the idea of "conceptual gateways" or "portals" through which learners are led towards what may previously have been (perceived) as a difficult or inaccessible way of thinking about something. This could be a new interpretation, understanding, or a change in worldview (p. 373). Ambiguity and uncertainty, as threshold concepts embedded in design studio education become, for students, a routine part of the learning process in design (Osmond, 2009).

Ambiguity, discussed as an interpretative relationship between people and artefacts in design has been distinguished into three types: ambiguity of *information*; of *context*; and of *relationship*. In the first, ambiguity refers to the way information is presented (in an artefact); in the second, it refers to different interpretations because of different contexts; and in the third, ambiguity is in the way a person relates to/feels towards an artefact (Gaver et al., 2003), drawing attention to the role of experience in relation to interpretation. Viewed through a semiotic lens, ambiguity is also discussed as an aesthetic function that focuses a person's attention and urges an interpretive effort while at the same time suggesting how to go about interpreting it. For (Eco, 1976), an ambiguous text operates through "a mode of violating the rules of the code" (p. 263), linking concepts not ordinarily associated to open up and amplify meaning, rather than being opaque and confusing (Phillips & McQuarrie, 2004). Ambiguity, metaphor, and paradox are all examples of terms that pertain to the production and exchange of meaning, and used in design, they draw attention to the role of language in the design process. A metaphor operates through substitution and is created through a conjunction of concepts, by stating or understanding one thing as another thing through a *comparison* which appears appropriate even though it may not be expected. A paradox also brings concepts together and is created through a *contradiction* which appears irregular but may reveal a truth. Both serve to draw a relation between how something is understood in the terms of something else. Where metaphor operates by substitution and comparison, and paradox by contradiction, ambiguity is created through *openness*. Despite the limitations of a structural approach, such interpretations of ambiguity are useful; they highlight that while ambiguous "texts" (here, social constructions) may have multiple meanings and be understood differently, as well as intersecting with many other texts, there is no error in the interpretation. In these ways, ambiguity is productive and positive; it implies an invitation to question and discover, urging towards more than one way of interpreting something, and is indicative of art and design's orientation to possibility and opportunity. These aspects are what make ambiguity in design education a productive force to support learners being led towards interpreting or understanding something in a new or different way, akin to a conceptual gateway.

At the same time, ambiguity runs counter to the quest for clarity, transparency, and certainty in higher education. In *Art and Design Pedagogy in Higher Education* (Orr & Shreeve, 2018) point to the perception of clarity in higher education as a "force for good" (p. 59), and of language (written) as transparent and unequivocal. This can create a tension in higher education and is in direct opposition to ambiguity as a force for change. However, language is not fixed, and as discussed above, as a social practice language shapes and is shaped by its culture of use; it is simultaneously a way of understanding and creating the world, a way of acting upon the world and each other, and also a means of transforming these operations. Culture is concerned with creating meaning, and also, following Foucault (2005) with the organisation and prioritisation of values with respect to those meanings. Here, culture is a set of values that are "accessible to everyone, but which at the same time gives rise to a mechanism of selection and exclusion" (p. 179). In this way a learning situation can be conceived as being circumscribed in part by the values of the educators and students, in which for each, some values are instilled with more importance, and some less so (or not at all)—whether or not this prioritisation of values is explicitly acknowledged. Conceiving of culture in this way is useful because it brings into consideration the ways in which an individual's degree of comfort with ambiguity is culturally specific and that learning situations are always context-bound (Visser & Visser, 2004).

For educators, learning as cultural and contextual in turn points to the importance of critically questioning one's own teaching practices, and what is conceived as valuable in approaching student learning. For example, if comfort with ambiguity is specific to culture, then ambiguity may take many forms and be experienced differently. The problem arises where there is a "discourse of acceptance of diverse outcomes but beneath is a hidden curriculum open only to a privileged few" (Austerlitz et al., 2008, p. 22). Working with ambiguity then as an approach to teaching and learning must be intentional, made explicit to students, and not make generalisations about how students respond.

This leads to what Orr and Shreeve (2018) refer to in art and design education as "a position of productive ambiguity" through which educators can consider ways to help students transition to higher education, and towards embracing ambiguity in preparation for uncertainties of future work. They advocate for the establishment of safe learning spaces, support for risk taking, and scaffolding forays into the unknown. And what may seem obvious, but may not always be made explicit to students talking with students about ambiguity, and about the individual discomforts that they may experience (p. 63). In tandem with these provisions is a framework for learning and interaction. These are the fixed components where there is no advantage to ambiguity in educational settings: course dates and structures, start and finish times of classes, information that is available about assignments across the various platforms is consistent. Importantly, to be aware of privileging particular curricula, histories, and knowledges which function to exclude as "ambiguity is not to be conflated with elitism" (p. 62). This is key as it points to the ethical and equitable treatment of ambiguity in teaching and learning. Educators must be acutely cognisant of the relations of power concerning ambiguity and their own positions where they are delivering and/or designing curriculum and learning activities, and assessing student work. It also demands that educators question the implicit knowledges embedded in their teaching practices, which while hidden, can operate to assimilate cultural specificity particularly into colonial structures that deny diverse approaches to learning and knowledge, and individual degrees of comfort with ambiguity.

Embracing a Pedagogy of Ambiguity

Ambiguity as a positive and productive force for change implies openness to alternative ideas and interpretations, whether it be information, a situation, or the way in which one feels towards something. To "embrace ambiguity" is to recognise its importance as part of everyday life, as central to learning and to the uncertainty of contemporary work (Austerlitz et al., 2008; O'Connor et al., 2018; Orr & Shreeve, 2018; Richardson, 2016; Suzawa, 2013; Visser & Visser, 2004). Austerlitz et al. (2008), in illustrating the value of art and design education, describe a *pedagogy of ambiguity* as approaches to teaching and learning "where skills are not simply competencies, but the ability to operate in the complexities of uncertainty" and a way to enhance students' opportunities in future work contexts (p. 6).

So, what does embracing a "pedagogy of ambiguity" mean for design education now, and for teaching and learning more broadly? It is important to consider curriculum, conceived here as interwoven with pedagogy and teaching (Leask, 2015). Beyond the content or formal syllabus, curriculum involves what is *actually* taking place, as well as the plan for what is to take place. In this way, curriculum refers to what is being learnt, why, how, when, and where it is being learnt, and the evidence that learning is taking place (Hicks, 2018). Where many of these aspects may appear evident in programme/course design and delivery, it's the hidden curriculum that requires attention when approaching ambiguity in teaching and learning. Through unintended and implicit messages, the hidden curriculum privileges certain positions, whose and what knowledge counts, conceals gender and cultural biases, and unstated intentions in learning (Hicks, 2018; Leask, 2015). And because the curriculum is constructed and communicated through language use, e.g., via dialogue and in written/visual/digital form, without interrogation such aspects can remain hidden.

As a function of learning in the design process, ambiguity provides students with spaces for discovery, interpretation, and possibility. Ambiguity functions as a kind of invitation towards interpretation, whether it be of information, context, or how a person feels towards something. Ambiguity is optimised through studio learning, particularly through dialogue and interaction between learners and the educator where multiple approaches to the same project are explored and students see each other's interpretations in progress. This facilitates ambiguity's function as a way of extending a learner's understanding and knowledge because it encourages the capacity to embrace competing ideas and multiple viewpoints. For design educators, embracing ambiguity enables space for studio learning to be imagined through curricula that expose hidden biases, privileged pedagogies, and what/whose knowledge counts.

The implicit structure of a signature pedagogy refers to the hidden curriculum: the tacit beliefs about professional values, dispositions, and attitudes (Shulman, 2005b), that are produced and reproduced through discourse. As any learning situation is also in part circumscribed by the values of the educators and the students, ambiguity is relational and contextual. For design educators, interrogating the hidden curriculum and continued utility of certain routinised practices and tacit beliefs about design will support the development of globally capable graduates able to navigate increasingly complex and uncertain professional contexts. And as ambiguity is tied to discovery and possibility, where ideas go beyond the anticipated, it "decentres the expertise of the teacher" (Orr & Shreeve, 2018, p. 64) which, when embraced, supports more reflexive teaching practices.

As a function of learning more broadly, embracing ambiguity affords educators opportunities to extend their knowledge and transform understanding across disciplines by facilitating "boundary crossings where closure between knowledge domains or areas of strictly defined expertise reigned before" (Nowotny, 2016, xi). Leask (2015) proposes an *Imagine* stage in constructing a curriculum where educators embrace ambiguity in coming together to interrogate the foundations of disciplinary knowledges. Ambiguity provides space for "challenging the traditional and the taken-for-granted", leading to new ways of interpreting student learning in curriculum (pp. 42–43). This approach to curriculum highlights an important aspect of ambiguity for teaching and learning; that its deployment by educators together as a purposeful activity supports expanding knowledge towards new possibilities for teaching and learning. In this regard, embracing ambiguity in teaching and learning means questioning embedded and tacit knowledge and the discourses through which this knowledge is reproduced and constrained.

To this end, avoiding ambiguity in higher education is not desirable; the ability to cross boundaries—disciplinary and contextual—to interpret and understand in different ways, is central to higher education's contribution to the social construction of individuals and learning as transformative—for both educators and students. At the same time, ambiguity in teaching and learning can create tensions. The quest for clarity and certainty in higher education runs counter to the need for ambiguity as a productive force for change. Some disciplines are also absolutely reliant on precision and certainty, e.g., safety concerns in bridge construction or clinical drug trials. However, as Nowotny (2016) points out "in the real world, things can become messy very quickly. Distinctions between acceptable risks and those deemed unacceptable are not easy to make" (xii). This is where ambiguity as an active process of learning has a broader role in preparing students for the complexity and uncertainty of work. Additionally, understanding of ambiguity is culturally and contextually specific. For educators, the idea of embracing ambiguity may spark tensions and discomfort around interrogating disciplinary knowledge and routinised practices. For students, tensions may arise where ambiguity is present in the curriculum however the rationale for it is not articulated or made explicit as a function of learning, potentially increasing anxiety, and negative attitudes to ambiguity. In sum, an intentional treatment of ambiguity in learning is needed to support students to decipher uncertainty and complexity in a future of work beyond known limits and routines.

Building on the research in design education and viewed through a lens of ambiguity as discourse, embracing a pedagogy of ambiguity brings into play certain provocations or triggers as starting points for exploration by educators involved in teaching and learning in higher education.

Ten Provocations for a Discussion on Ambiguity for Teaching and Learning in Higher Education

1. **The tension between ambiguity and the pursuit of clarity in higher education**

Maximising clarity and certainty are at the heart of risk management strategies in higher education. This can run counter to the role of ambiguity as a productive force for change. How might we reconcile the two in order to best support student learning?

2. **Holding competing ideas and multiple perspectives**

Ambiguity is embraced in design studio learning where students are encouraged to see multiple, tangible interpretations of the same thing and embrace competing ideas. How might we promote this approach in non-studio-based learning environments?

3. **Discourse privileges certain knowledge and values**

Implicit and tacit values and knowledge are part of any learning environment and are established through discourse. Different discourses circumscribe and afford different positions for educators and learners that can exclude alternative knowledges, histories, and values and reduce the capacity of learners to work with ambiguity as a productive and creative force. How do we create safe and equitable spaces to explore the hidden curriculum and approaches to the ambiguity that are respectful and inclusive?

4. **Ambiguity is relational and contextual**

Ambiguity is socially constructed in its context and culture of use; it can take many forms and be experienced differently. To support student learning, ambiguity must be considered in relation to all the aspects of a teaching and learning situation where curriculum, pedagogy, and teaching are interwoven. What can be done to optimise ambiguity in learning without problematising the individual, or making generalisations about students?

5. **Ambiguity in learning must be intentional and explicit**

To support student learning, an approach to ambiguity must be intentional and explicit rather than haphazard or random. It must differentiate between those aspects of learning that are fixed (e.g., course, class, assessment dates and times, syllabus information, course objectives) and the processes of learning that are open to competing ideas and a multiplicity of interpretations.

6. **Ambiguity provides space for discovery**

Talk with students from the outset of and throughout their overall programme/course about ambiguity as central to learning, and to the ability to interpret and operate in the complexities and uncertainties of work.

7. Physical studio learning is not necessarily inclusive

Studio learning embraces ambiguity because it aspires to encourage everyone to see each other's work. What the global pandemic and online learning have demonstrated is an enhanced capacity to create space for a diverse range of student voices that aren't always heard in a traditional studio. How might we understand ambiguity in studio learning as not contingent on the physical environment?

8. Signature pedagogies evolve over time

Signature pedagogies are pervasive, habitual, and routinised ways of learning and doing that enable learners to focus on the increasingly complex subject matter as they become enculturated into a profession and community of practice. Because the world and the nature of work are changing, the utility of signature pedagogies must be interrogated.

9. There is a right place for ambiguity in all learning

Not all disciplines may believe they have the luxury of embracing ambiguity. However, ambiguity as an active function of learning and interpretation has a broader role in preparing students for the uncertainty of future work. How do we reconcile ambiguity as a process of learning with the need for certainty and precision in a discipline?

10. Ambiguity enables boundary crossing

A mindset of deep engagement with ambiguity creates spaces for "challenging the traditional and taken-for-granted" in disciplinary knowledge (Leask, 2015). How might educators come together and use this mindset to design across disciplines?

Conclusion

Higher education is increasingly complex, changing, and characterised by ambiguity, and the outlook for the labour market is highly uncertain. Embracing a pedagogy of ambiguity in higher education is proposed as a way of supporting students to prepare for the uncertain world of work, and for educators to interrogate embedded knowledge towards new possibilities for teaching and learning in design and more broadly. Ambiguity is productive and positive, a force for change that is part of everyday life and integral to learning. Because ambiguity, as a discourse, is culturally and socially constructed in its context of use, it is not my intent to provide a single framework or toolkit for teaching ambiguity. Rather, the chapter concludes with a series of provocations as starting points for educators to explore their own teaching practices in purposefully designing for ambiguity. There is scope for further work in exploring these and other provocations across disciplines and universities to increase understanding of the cultural and contextual specificities of ambiguity as a productive force that supports learning and is part of everyday life.

References

Austerlitz, N., Blythman, M., Jones, B. A., Jones, C. A., Grove-White, A., Morgan, S. J., Orr, S., & Shreeve, A. (2008). Mind the gap: Expectations, ambiguity and pedagogy within art and design higher education. In L. Drew (Ed.), *The student experience in art and design higher education: Drivers for change* (pp. 125–148). Jill Rogers Associates.

Bolman, L. G., & Gallos, J. V. (2021). *Reframing academic leadership* (2nd ed.). Jossey-Bass.

Brown, J. B. (2020). From denial to acceptance: A turning point for design studio in architecture education. *Distance Design Education*. https://distancedesigneducation.com/2020/05/11/from-denial-to-acceptance-a-turning-point-for-design-studio-in-architecture-education/

Budner, S. (1962). Intolerance of ambiguity as a personality variable. *Journal of Personality, 30*(1), 29–50.

Buller, J. L. (2014). *Change leadership in higher education a practical guide to academic transformation*. Wiley.

Dunican, B., Gearin, C. A., & Norman, T. (2019). Exploring resistance to change and intolerance to ambiguity in higher education institutions. *International Journal of Leadership and Change, 7*(1), 41–47.

Eco, U. (1976). *A theory of semiotics*. Indiana University Press.

Fairclough, N. (1993). *Discourse and social change*. Polity Press.

Fairclough, N. (2002). Critical discourse analysis as a method in social scientific research. In R. Wodak & M. Meyer (Eds.), *Methods of critical discourse analysis* (pp. 121–138). SAGE Publications.

Fleischmann, K. (2021). Is the design studio dead? An international perspective on the changing shape of the physical studio across design domains. *Design and Technology Education: An International Journal, 26*(4), 112–129.

Foucault, M. (2005). *The hermeneutics of the subject: Lectures at the Collège de France 1981–1982* (G. Burchell, Trans.; F. Gros, F. Ewald, & A. Fontana, Eds.). Picador.

Frenkel-Brunswick, E. (1949). Intolerance of ambiguity as an emotional and perceptual personality variable. *Journal of Personality, 18*, 108–143.

Furnham, A., & Marks, J. (2013). Tolerance of ambiguity: A review of the recent literature. *Psychology, 04*, 717–728.

Gaver, W., Beaver, J., & Benford, S. (2003). *Ambiguity as a resource for design CHI 2003.* Human Factors in Computing Systems.

Gray, C. M. (2021). "Scaling up" and adapting to crisis: Shifting a residential UX Studio Program Online. *Design and Technology Education: An International Journal, 26*(4), 37–49.

Herman, J. L., Stevens, M. J., Bird, A., Mendenhall, M., & Oddou, G. (2010). The tolerance for ambiguity scale: Towards a more refined measure for international management research. *International Journal of Intercultural Relations, 34*(1), 58–65.

Hicks, O. (2018). Curriculum in higher education: Confusion, complexity and currency. *HERDSA Review of Higher Education, 5*, 5–30.

Jones, D. (2020). Creating distance design courses: A guide for educators. https://distancedesigneducation.com/creating-distance-design-courses/.

Jones, D., & Lotz, N. (2021). Design education: Teaching in crisis. *Design and Technology Education: An International Journal, 26*(4), 4–9.

Kelly, V., Moline, K., & Vaughan, L. (2021). Design education now. *NiTRO: Non Traditional Research Outcomes, 36*. https://nitro.edu.au/articles/2021/7/30/design-education-now

Leask, B. (2015). *Internationalizing the curriculum.* Routledge.

Lehtonen, M. J., Khamisani, N., & Gatto, G. (2021). Playful absence/absence of play. *Design and Technology Education: An International Journal, 26*(4), 20–36.

Mahmoud, N., Kamel, S., & Hamza, T. (2020). The relationship between tolerance of ambiguity and creativity in architectural design studio. *Creativity Studies, 13*, 179–198.

Marshall, S., Blackley, S., & Green, W. (2022). 40 years of research and development in higher education: Responding to complexity and ambiguity. *Higher Education Research & Development, 41*(1), 1–6.

Marshalsey, L., & Sclater, M. (2020). Together but apart: Creating and supporting online learning communities in an era of distributed studio education. *International Journal of Art & Design Education, 39*(4), 826–840.

McLain, D. L. (2009). Evidence of the properties of an ambiguity tolerance measure: The multiple stimulus types ambiguity tolerance scale—II (MSTAT—II). *Psychological Reports, 105*(3), 975–988.

Meyer, J. H. F., & Land, R. (2005). Threshold concepts and troublesome knowledge (2): Epistemological considerations and a conceptual framework for teaching and learning. *Higher Education, 49*, 373–388.

Mintz, B. (2021). Neoliberalism and the crisis in higher education: The cost of ideology. *The American Journal of Economics and Sociology, 80*(1), 79–112.

Norton, R. W. (1975). Measurement of ambiguity tolerance. *Journal of Personality Assessment, 39*(6), 607.

Nowotny, H. (2016). *The cunning of uncertainty*. Polity Press.

O'Connor, P., Becker, K., & Fewster, K. (2018). Tolerance of ambiguity at work predicts leadership, job performance, and creativity. In: *Creating uncertainty conference*, Ascona, Switzerland.

Orr, S., & Shreeve, A. (2018). *Art and design pedagogy in higher education: Knowledge, values and ambiguity in the creative curriculum*. Routledge.

Osmond, J. (2009). 'Stuck in the bubble': Identifying threshold concepts in design. In D. Clews (Ed.), *Dialogues in art & design: Promoting and sharing excellence* (pp. 130–135). ADM-HEA/GLAD.

Osmond, J., & Turner, A. (2010). The threshold concept journey in design: From identity to application. In J. H. F. Meyer, R. Land, & C. Baillie (Eds.), *Threshold concepts and transformational learning* (Vol. 42, pp. 347–363). Brill.

Phillips, B. J., & McQuarrie, E. F. (2004). Beyond visual metaphor: A new typology of visual rhetoric in advertising. *Marketing Theory, 4*(1/2), 113–136.

Richardson, D. F. (2016). *Toward a pedagogy of ambiguity: Incorporating and assessing ambiguity in a multiliteracies based foreign language classroom*. Doctoral Thesis, University of Arizona.

Shreeve, A. (2012). Signature pedagogies in design: Linking teaching, learning and practice. *Art+Design\Education: Collection, 4*(Spring), 43–51.

Shreeve, A. (2015). Signature pedagogies in design. In M. Tovey (Ed.), *Design pedagogy: Developments in art and design education* (pp. 83–94). Gower Publishing.

Shulman, L. S. (2005a). Pedagogies of uncertainty. *Liberal Education, 91*(2), 18–25.

Shulman, L. S. (2005b). Signature pedagogies in the professions. *Daedalus, 134*(3), 52–59.

Suzawa, G. S. (2013). The learning teacher: Role of ambiguity in education. *Journal of Pedagogy, 4*(2), 220–236.

Visser, J., & Visser, Y. (2004). Ambiguity, cognition, learning, teaching, and design. *TechTrends, 48*, 40–43.

World Economic Forum. (2020). *The future of jobs report 2020*. World Economic Forum.

CHAPTER 6

Designing for X-Disciplinarity: Why, What and How? Why Now?

JiaYing Chew

When one says "design(ing) across disciplines", several words immediately come to mind: Multi-/Inter-/Cross-/Transdisciplinary. Often, these terms are vaguely defined and used interchangeably (Gasper, 2010; Jahn et al., 2012). It is difficult to come to a consensus on how these initiatives should be enacted in teaching and learning practice, but the underlying premise is one and the same—for those participating in these activities to have a more holistic understanding of the ever-changing societal challenges at hand.

Despite the litany of the world's challenges that calls for insights from disparate fields—climate change, poverty, world hunger, disease—it

J. Chew (✉)
Division of Industrial Design, National University of Singapore, Singapore, Singapore
e-mail: jiaying.chew@nus.edu.sg

London College of Communication, University of the Arts London, London, United Kingdom

M. J. Lehtonen et al. (eds.), *Design Education Across Disciplines*,
https://doi.org/10.1007/978-3-031-23152-0_6

is an open secret that designing across disciplines within higher education institutions (HEI) is challenging for varied reasons (Gibbs, 2017; Klein, 1996). However, it is safe to say that the Covid-19 pandemic demonstrated how the social problems confronting society today are unprecedented and require specialised knowledge from disparate fields that needs to be brought to bear on the issue at hand (Jacobs, 2014). Through the effects of the pandemic, we experienced globally what Taylor (2010, pp. 151–152 in Jacobs, 2014) described of the diabetes epidemic, that "the causes are not only medical but also social, political, economic, environmental and psychological as well". The pandemic has no doubt acted as a catalyst for higher education leadership to acknowledge the dire urgency and necessity to provide a top-down push for and tangible support of interdisciplinary endeavours. Despite the different strategies, we see how they have been acted upon as observed through the examples of The National University of Singapore (National University of Singapore, 2021), King's College London (Thain, 2021) and TD School at the University of Technology Sydney (UTS, 2021), for instance.

Thus, going beyond the discord of labels and conceptual overlaps, it is perhaps more useful to navigate this topic of crossing disciplines from strategic (why are we doing it and what is the vision?); tactical (what are the possibilities, connections and processes?) and operational (how can we approach it tangibly?) perspectives (Best, 2006, p. 17) to get a practical sense of what this means for us in higher education. As such, the chapters in Part I—"Higher education leadership and program management perspective" put forth insightful and, at times, provocative thoughts in this context. Before engaging with the chapters in more detail, let us first turn to the origins of transdisciplinarity and how the initial discussions in the 1970s are relevant more than ever today.

The transdisciplinary agenda for developing HEI was first proposed in the 1970s by the Organisation for Economic Co-operation and Development (OECD) (CERI, 1972, see also Jantsch, 1972; Piaget, 1972). In the decades that followed, we have witnessed an increase in inter- and transdisciplinary research amongst the OECD member countries (Klein, 1996; Nicolescu, 1999). There is a growing body of research dealing with transdisciplinarity and more recently transdisciplinarity has also been expanded to look at how research can contribute to solving complex (or wicked, as per Rittel & Webber, 1973) environmental and social problems. Similarly, educational initiatives going beyond single disciplines have also been gaining momentum (Nicolescu, 2005; Russell et al., 2008) as a response to tackling wicked problems (Buchanan, 1992; Laasch et al., 2020) as well as acknowledging the disciplinary limitations (Bremner & Rodgers, 2013; Brown & Katz, 2011). Discussions revolving around the transdisciplinary agenda have surfaced in recurrent cycles over the decades since their origins five decades ago, yet there seems to be no clear direction nor alignment on how it should be approached despite growing demands and recognition.

Positioning

Before we continue, I would like to address my positioning and stance when synthesising the chapters. I am a multi-disciplinarian by training—a designer with a background in facilities management, one who treads the peripheries of "design", dabbling in education and management. As I am writing this synthesis, I am an Education Design Strategist at the National University of Singapore, and a Ph.D. researcher at the University of the Arts London (London College of Communication). This topic of crossing boundaries first piqued my interests when I was a student of an interdisciplinary graduate programme, where I developed a keen interest in understanding how the teaching and learning model for interdisciplinarity could be improved and refined to better enable knowledge exchange and integration amongst both students and faculty (Chew et al., 2020). Since then, I have been exploring the creation of tools that enable a common platform for faculty members of different disciplines to co-design for transdisciplinarity, which I believe is a fundamental aspect to overcoming the teaching challenges within cross-disciplinary initiatives (Chew, 2021). My current research looks at the role of design in transdisciplinary higher education, for which I am studying how design can be leveraged as a common language (Bremner & Rodgers, 2013) and shared logic (Ertas et al., 2003; Garbuio et al., 2018; Jahn et al., 2012) in curriculum development for teaching and learning, but also understanding how design enquiry can be a tool to encourage re-evaluation of existing assumptions and encourage collaboration (Lee et al., 2018).

Thus, I approach these chapters through the lens of a design educator and action researcher that is designing, developing, and delivering an undergraduate transdisciplinary pilot module that leverages on the design studio pedagogy (Shulman, 2005). Having experienced first-hand, the challenges of crossing disciplinary boundaries within a research university; I must stress that collaboration and communication are vital to bridge the varying epistemic frameworks when different disciplines are brought together. There are multiple layers to peel apart when we discuss how to approach inter- or transdisciplinary studies because it is not as simple as putting different disciplines at the same table, rather, it all starts with how we approach teaching in the classroom. Bringing together different disciplines is comparably easier when funded research is involved. However, when it comes to navigating new modes of team teaching in a diverse setting; relinquishing control and tradition from one's own discipline seems to be a tedious challenge. Furthermore, even when "the table is set", it requires conscious effort from all involved parties to design and orchestrate how the disciplines should be weaved together coherently—the ideology, methods and space are all part of the equation to the successful delivery of a transdisciplinary curriculum. Especially because the way individuals understand concepts contains traces of disciplinary specificities; without clarity through a common platform, the messiness is reflected in the curriculum design, development and delivery of interdisciplinary courses (Chew, 2021). Therefore, the "Why", "What" and "How" we might approach the topic is key to creating a solid baseline of understanding, otherwise we will fall into a vicious cycle of this "x-disciplinary"[1] topic fizzling out again—as we have seen in the past when interdisciplinary activities were undermined by "misinformation, bias and easy generalisations", especially when funding priorities had shifted (Klein, 1996).

Challenges for Transdisciplinarity

It is worth mentioning that initially the transdisciplinary agenda was conceived as a response to addressing the rigidity of disciplinary silos: "to meddle with the disciplines is to meddle with the social structure of the university in its entirety" (CERI, 1972, p. 9). While the initial agenda was transported to addressing societal and environmental issues through research (Brandt et al., 2013; Max-Neef, 2005), during the twenty-first century we are still witnessing debates concerning the future of universities (Laasch et al., 2020; Tan, 2020) that seem to echo the initial conversations in the 1970s (Jantsch, 1972; Piaget, 1972; Rittel &

[1] Here, the term "x-disciplinary" is intentional and intended to demonstrate the dropping of labels for crossing of disciplinary boundaries as the labels sometimes creates more confusion than clarity.

Webber, 1973) as we begin to acknowledge the need to teach and learn in transdisciplinary ways.

Nonetheless, it is equally important to acknowledge that universities were initially conceived to develop a compliant workforce with skills of print literacy and discipline that would enable them to function in modern corporations and a corporate economy (Kellner, 2003). The University then subsequently became a "mechanism" for assuring society that only the qualified would be allowed to practise certain professions (Collini, 2012), and is also the traditional institution most responsible for making constant societal changes in human history (Seymour & Fife, 1988). However, it is no longer sufficient for universities to remain merely as sites of knowledge production and dissemination (Gunasekara, 2004; Russell et al., 2008). Universities have been pressured to utilise their capacities in applying and lending their knowledge and strengths to economic growth, as well as finding solutions to industrial and societal challenges (Collini, 2012; Russell et al., 2008; Sutz, 1997a, 1997b). As a response, and often going beyond pure economic gains, educators have explored ways through which students could approach such challenges from diverse perspectives (e.g., Lehtonen et al., 2022).

Despite interest from faculty members and a push from industry; "x-disciplinarity" remains an elusive ambition for universities (Klein, 2009) as most institutions are still not agile enough to embrace and tackle the operational challenges that emerge from inter- and transdisciplinary educational needs. Furthermore, as Graff (2016) describes, there is little to no consensus on what the term "Interdisciplinarity" entails, much less an agreement of how it should be approached especially in teaching and learning. To some, interdisciplinary work is a badge of honour; yet to others, it is damning, as it stimulates fears that when unchecked, interdisciplinarity will undo the legitimate and recognised traditions of checks, balances and faculty authority (Jacobs, 2014). However, the deepening of individual professions to ensure this "qualification" has also resulted in the increasingly siloed operations of the institution. So the question persists: how are we to solve the pressing issues of the larger societal context if we are seeing but one facet of a bigger picture?

Different Parts of the Same Equation

Thus, in this collection of chapters, putting aside confusing labels of x-disciplinary studies, we focus on the practical notions of crossing disciplinary boundaries instead, and why some key factors set apart design's signature pedagogies (McLain, 2021; Shulman, 2005) might be worth considering for those embarking on x-disciplinary endeavours. Authors of each chapter have provided compelling viewpoints that hopefully spark further conversations and concrete initiatives on how we might envision pathways for achieving the transdisciplinary agenda from the 1970s and, eventually, go beyond that. It must be pointed out, however, that the transdisciplinary agenda ought not to be taken as a project for dissolving the disciplines. This means that the term transdisciplinarity should not be seen as synonymous with the terms "anti- or uni-disciplinary", terms that have connotations of academic borders being unnecessary. Rather, the advocacy is that disciplines have served and continue to serve knowledge organising purposes but they should not be treated as the only lenses through which we approach the world (Russell, 2005).

Providing an overview to one of the key challenges in x-disciplinary teaching and learning, Kelly's chapter discusses the notion of uncertainty and ambiguity as key themes to navigating the future of work and higher education. In it, the author acknowledges the tension between academia's pursuit of clarity that possibly impedes the need for ambiguity as a productive force for change as we are increasingly required to tackle complex social issues that involve stakeholders with multifarious agendas (Buchanan, 1992). Kelly highlights the relevance of design's signature pedagogies (Shulman, 2005) where ambiguity and uncertainty are central concepts to the classroom, particularly in studio-based learning where students engage directly with the complexity of the subject matter through project-based learning rather than trying to conform to the right answers against a rubric. The design studio provides a safe space for exploration by employing the idea of a "threshold concept" (Meyer & Land, 2005), for which learners are provided a basic scaffolding of a topic that might be perceived as difficult. This eases them into the process of navigating a seemingly overwhelming challenge, thus giving them the agency to take charge of their own discoveries to develop creative solutions in response to complex issues (Lehtonen et al., 2022). Leveraging on the attributes of the design studio could potentially provide a baseline to support transitions in higher education to encompass greater levels of ambiguity, allowing both students and faculty across disciplines to navigate the unknown together.

Further supporting the proponent of the design studio, Rowe delves deeper into a non-exhaustive list of characteristics unique to design pedagogy or Design-Based Learning (Davis, 1998), featuring collaborative learning; problem-focused scenarios; and taking a future-oriented perspective, just to name a few. This chapter opens a discussion around each of the characteristics, providing a practical explanation of the way design studios are set up (McLaughlan & Lodge, 2018) and how each of the listed characteristics manifests in the approach we take to teaching and learning in design. The sum of these characteristics in turn creates a safe space that allows for "constructive failure", a mindset not commonly embraced in higher education. Yet, in a future that sees constant change and continually shifting landscapes, higher education will need to remodel itself as educators shift from being knowledge providers to facilitators (Badley & Habeshaw, 1991). Thus, Rowe suggests that some of these pedagogic practices, though familiar to design, may be new and cutting edge to the rest of the university and adopting some of these traits from design pedagogy could be potentially useful in the transformations required.

Demonstrating how the design studio can function as a vehicle for students to engage in complex social issues beyond the classroom, Brophy et al.'s chapter takes the reader through the setup and intent of a first-year transdisciplinary design unit in an Australian University. Through this design unit, design students are exposed to transdisciplinary learning through collaborative projects with external industry and community organisations to engage in complex social issues. In this example, the authors bring attention to the fact that even under the broad umbrella of design, there seem to be silos and that these can be bridged by bringing together the different sub-disciplines of design under one design studio where students work on socially responsive design projects. By leveraging on design's signature pedagogy, students learn by doing and by giving students a sense of agency. Through this they learn how to navigate the ambiguity of wicked problems with the contemporary examples of societal challenges as a threshold concept (Meyer & Land, 2005).

Expanding upon the definition of transdisciplinary that goes beyond design-based disciplines, Keane & Yeow's chapter takes a systems-level approach to deconstructing and re-constructing teaching curricula and pedagogies at various levels in higher education institutions. Through key exemplars in the United Kingdom, the authors describe how universities might consider remodelling to enable smoother transitions that allow students to be better prepared for the workforce. First, within the classroom: Taking a student-centred approach to interdisciplinary teaching and learning, complex problems, and problem-solving and act as the centrepiece for a collaboration between a business school and an art school. By drawing on the respective strengths of the disciplines involved, students can engage in a deep exchange of disciplinary perspectives and enhance their professional practice. Second, taking the learning outside of the classroom: Offering hands-on, experiential learning through degree apprenticeship, and internships allows for better integration of industry partners into the university ecosystem and for students to learn through doing (Lave & Wegner, 1991), to better understand the needs of industry and society. Third, creating an inclusive third space: Establishing Interdisciplinary Centres of Excellence to encourage and enable academic teams to work across disciplinary boundaries to provide solutions to complex problems. This allows for collaborations between two or more subject specialisms, providing a platform to bridge perspectives and create opportunities to start important conversations. Breaking down the silos requires taking a holistic, systems approach in the university ecosystem. This allows us to reimagine the institution as one that is agile enough to keep up with contemporary needs.

X-Disciplinarity: Looking Ahead

Illustrated in Fig. 6.1 is my take on how each chapter in this section contributes to an ongoing and continually unfolding discussion about how higher education needs to be more progressive and adaptive to embrace x-disciplinarity.

To synthesise, the chapters in Part I suggest how design's signature pedagogies (Shulman, 2005) could be an instrumental piece to aid the much-needed transformations of higher education institutions becoming future-ready, as ambiguity and uncertainty are central themes in the world's current problems. In design pedagogy, students are dealt with design challenges and expected to develop future-oriented solutions in a collaborative manner. The design approach thus creates a common platform or "threshold concept" (Meyer & Land, 2005) that allows for an entry point to understanding complex and multi-faceted societal issues rather than expecting an immediate solution. Moreover, the iterative

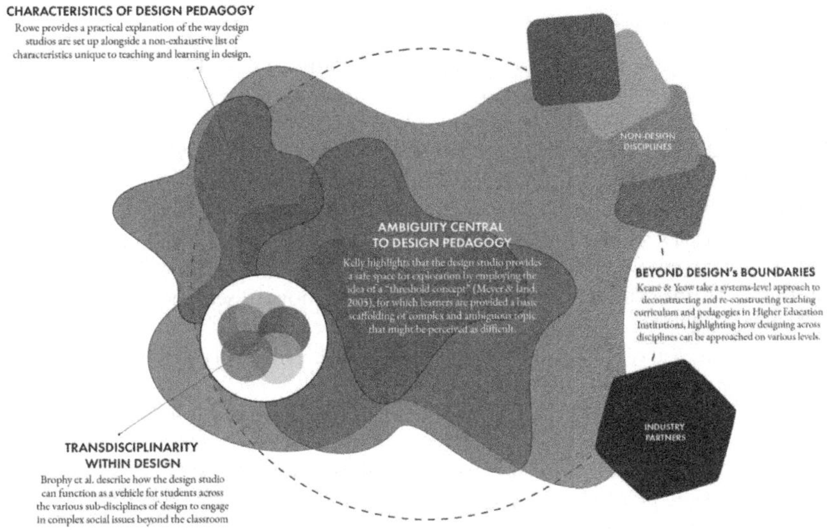

Fig. 6.1 Synthesis of chapters in Part I

nature of design (McLaughlan et al., 2021) also allows students to experiment and fail in a safe space, as opposed to being expected to know the right answers in an examination. Thus, the uniqueness of design pedagogy enables students to be equipped with the mindset and attitudes necessary to approach seemingly impossible challenges, which non-design disciplines could potentially benefit from as well.

Finally, as we have observed, the operational challenges of "x-disciplinarity" persist now, despite the origins of the topic since the 1970s. Designing across disciplines will thus require an inclusive roadmap that both respects the need for disciplinary depth and knowledge, while also navigating the newness and unchartered waters. In the multi-layered approach to organisational transformations, it seems that the approach in design pedagogy has much to offer when extrapolated to other parts of higher education and ideally will spark conversations that enable smoother transitions and innovations within universities that we can hopefully look forward to.

References

Badley, G., & Habeshaw, T. (1991). The Changing role of the teacher in higher education. *British Journal of in-Service Education, 17*(3), 212–218. https://doi.org/10.1080/0305763910170307

Best, K. (2006). *Design management: Managing design strategy, process and implementation*. AVA publishing.

Brandt, P., Ernst, A., Gralla, F., Luederitz, C., Lang, D. J., Newig, J., Reinert, F., Abson, D. J., & von Wehrden, H. (2013). A review of transdisciplinary research in sustainability science. *Ecological Economics, 92*, 1–15.

Bremner, C., & Rodgers, P. (2013). Design without discipline. *Design. Issues, 29*(3), 4–13.

Brown, T., & Katz, B. (2011). Change by design. *Journal of Product Innovation Management, 28*(3), 381–383.

Buchanan, R. (1992). Wicked problems in design thinking. *Design Issues, 8*(2), 5–21.

CERI (Centre for Educational Research and Innovation) (Ed.). (1972). *Interdisciplinarity problems of teaching and research in universities*. CERI.

Chew, J.-Y. (2021). Siloed in breaking silos: A case of interdisciplinary curriculum (mis)alignment. In: *Proceedings of the DRS learn X design 2021: 6th international conference for design education researchers* (Vol. 2, pp. 489–498). Design Research Society.

Chew, J.-Y., Lee, J., & Lehtonen, M. J. (2020). Towards design-driven transdisciplinary education: Navigating the challenges and envisioning the role of design as a facilitator. In S. Boess, M. Cheung, & R. Cain (Eds.), *Synergy—DRS international conference 2020, 11–14 August* (pp. 1847–1866). Design Research Society.

Collini, S. (2012). *What are universities for?* Penguin.

Davis, M. (1998). Making a case for design-based learning. *Arts Education Policy Review, 100*(2), 7–15.

Ertas, A., Maxwell, T., Rainey, V. P., & Tanik, M. M. (2003). Transformation of higher education: The transdisciplinary approach in engineering. *IEEE Transactions on Education, 46*(2), 289–295.

Garbuio, M., Dong, A., Lin, N., Tschang, T., & Lovallo, D. (2018). Demystifying the genius of entrepreneurship: How design cognition can help create the next generation of entrepreneurs. *Academy of Management Learning & Education, 17*(1), 41–61.

Gasper, D. (2010). Interdisciplinarity and Transdisciplinarity. Diverse purposes of research: Theory-oriented, situation-oriented, policy-oriented. In P. Thomson & M. Walker (Eds.), *The Routledge doctoral student's companion* (p. 444). Routledge.

Gibbs, P. (2017). *Transdisciplinary higher education: A theoretical basis revealed in practice*. Springer.

Graff, H. J. (2016). The "problem" of interdisciplinarity in theory, practice, and history. *Social Science History, 40*(4), 775–803.
Gunasekara, C. (2004). The third role of Australian universities in human capital formation. *Journal of Higher Education Policy and Management, 26*(3), 329–343.
Jacobs, J. A. (2014). *In defense of disciplines*. University of Chicago Press.
Jahn, T., Bergmann, M., & Keil, F. (2012). Transdisciplinarity: Between mainstreaming and marginalization. *Ecological Economics, 79*, 1–10.
Jantsch, E. (1972). Inter- and transdisciplinary university: A systems approach to education and innovation. *Higher Education, 1*(1), 7–37.
Kellner, D. (2003). Toward a critical theory of education. *Democracy & Nature, 9*(1), 51–64.
Klein, J. T. (1996). *Crossing boundaries: Knowledge, disciplinarities, and interdisciplinarities*. University of Virginia Press.
Klein, J. T. (2009). *Creating interdisciplinary campus cultures: A model for strength and sustainability*. John Wiley & Sons.
Lave, J., & Wenger, E. (1991). *Situated learning: Legitimate peripheral participation*. Cambridge University Press.
Lee, J.-J., Jaatinen, M., Salmi, A., Mattelmäki, T., Smeds, R., & Holopainen, M. (2018). Design choices framework for co-creation projects. *International Journal of Design, 12*(2), 15–31.
Lehtonen, M. J., Yeow, P., & Chew, J.-Y. (2022). Empowering change for future-making: Developing agency by framing wicked problems through design. *Futures, 139*, 102952.
Max-Neef, M. A. (2005). Foundations of transdisciplinarity. *Ecological Economics, 53*(1), 5–16.
McLain, M. (2021). Towards a signature pedagogy for design and technology education: A literature review. *International Journal of Technology and Design Education*.
McLaughlan, R., & Lodge, J. M. (2018). Facilitating epistemic fluency through design thinking: A strategy for the broader application of studio pedagogy within higher education. *Teaching in Higher Education, 24*(1), 81–97.
McLaughlan, R., Pert, A., & Lodge, J. M. (2021). Productive uncertainty: The pedagogical benefits of co-creating research in the design studio. *International Journal of Art & Design Education, 40*(1), 184–200.
Meyer, J. H. F., & Land, R. (2005). Threshold concepts and troublesome knowledge (2): Epistemological considerations and a conceptual framework for teaching and learning. *Higher Education, 49*(3), 373–388.
National University of Singapore. (2021). *Two new colleges at NUS to deliver flexible, interdisciplinary education more accessibly, and at greater scale*. Two New Colleges at NUS to Deliver Flexible, Interdisciplinary Education More

Accessibly, and at Greater Scale. https://news.nus.edu.sg/two-new-colleges-at-nus-to-deliver-flexible-interdisciplinary-education-more-accessibly-and-at-greater-scale/
Nicolescu, B. (1999). The transdisciplinary evolution of learning. In: *Symposium on overcoming the underdevelopment of learning at the annual meeting of the* American Educational Research Association, Montreal, Canada.
Nicolescu, B. (2005). Towards transdisciplinary education. *The Journal for Transdisciplinary Research in Southern Africa, 1*(1), 5–16.
Laasch, O., Moosmayer, D., Antonacopoulou, E., & Schaltegger, S. (2020). Constellations of transdisciplinary practices: A map and research agenda for the responsible management learning field. *Journal of Business Ethics, 162*(4), 735–757. https://doi.org/10.1007/s10551-020-04440-5
Piaget, J. (1972). *The epistemology of interdisciplinary relationships* (pp. 127–139). Problems of Teaching and Research in Universities.
Rittel, H. W. J., & Webber, M. M. (1973). Dilemmas in a general theory of planning. *Policy Sciences, 4*(2), 155–169.
Russell, A. W. (2005). No academic borders? Transdisciplinarity in university teaching and research. *Australian Universities' Review, 48*(1), 35–41.
Russell, A. W., Wickson, F., & Carew, A. L. (2008). Transdisciplinarity: Context, contradictions and capacity. *Futures, 40*(5), 460–472.
Self, J. A., & Baek, J. S. (2017). Interdisciplinarity in design education: Understanding the undergraduate student experience. *International Journal of Technology and Design Education, 27*(3), 459–480.
Seymour, D. T., & Fife, J. D. (1988). *Developing academic programs: The climate for innovation* (ASHE-ERIC Higher Education Report No. 3, 1988). ASHE-ERIC Higher Education Reports, The George Washington University, One Dupont Circle, Suite 630, Dept. https://eric.ed.gov/?id=ED305015
Shulman, L. S. (2005). Signature pedagogies in the professions. *Daedalus, 134*(3), 52–59.
Sutz, J. (1997a). The new role of the university in the productive sector. *Universities and the Global Knowledge Economy*, 11–20.
Sutz, J. (1997b). The new role of the university in the productive sector. In H. Etzkowitz & L. Leydesdorff (Eds.), *Universities and the global knowledge economy: A triple helix of university–industry relations* (pp. 11–20). Pinter.
Tan, E. C. (2020, September 10). Universities need to tear down subject silos. *The Straits Times*, p. 21.
Thain, M. (2021, July 18). Post-pandemic recovery requires radical interdisciplinarity. *Times Higher Education (THE)*. https://www.timeshighereducation.com/blog/post-pandemic-recovery-requires-radical-interdisciplinarity
UTS. (2021, July 28). *Introducing UTS's new TD school.* University of Technology Sydney. https://www.uts.edu.au/about/td-school/news/introducing-utss-new-td-school

PART II

Design Education Blending and Crossing Boundaries: Practices for the 21st Century

CHAPTER 7

Designing a Tactile Class in Online Learning: "Click-Points" vs Touchpoints

Aldo Valencia, Trevor Vaugh,
and Threase Finnegan-Kessie

Introduction

The fourth industrial revolution is meant to blur boundaries between the biological, physical, and digital worlds (Schwab, 2016). In this century, technology has become more and more integrated into our social interactions with other people and with nature. Recent medical and technological breakthroughs such as Neuralink and Braingate have given us a glimpse of how embedded technology within the human body and

A. Valencia (✉)
Design Innovation Department, Maynooth University, Maynooth, Ireland
e-mail: aldovalenciah@gmail.com

T. Vaugh
Design Innovation Department, Maynooth University, Maynooth, Ireland

T. Finnegan-Kessie
Design Innovation Department, Maynooth University, Maynooth University, Maynooth, Ireland

M. J. Lehtonen et al. (eds.), *Design Education Across Disciplines*,
https://doi.org/10.1007/978-3-031-23152-0_7

brain could change the trajectory of human civilisation. As technology is enhancing our outreach, capacities, and volume of interactions in Higher Education (HE), it is fair to say that it also separates the engagement between staff and students, lowering the attention span and motivation and, overall, impacting the student and staff experience.

Online learning is set to become a vital factor in the digital transformation of Higher Education institutions in the twenty-first century. However, this shift happened swiftly, much sooner than expected. Staff and students had to adapt to these changes while the university structure stayed in an uncertain state. As pointed out by Lehtonen et al. (2021), Covid-19 has brought up the opportunity to explore new ways of content delivery at university, creating a void space in the teaching and learning experience that could lead to the establishment of new conventions and ways of knowing and inquiring.

In this chapter, we describe the adaptation of multimodal learning activities to an online classroom to dodge the dullness and lethargy of ill-prepared virtual interaction. The participants were enrolled in the module called Design, Entrepreneurship, and Innovation, which is offered to second-year students of international business, marketing, media communication, arts, and science at Maynooth University. The aim of these activities was to use tactile elements during class to add an extra learning mode vaguely explored in other online learning modules. The activities planned for this study focused on the emergence of 3D tactile metaphors to spark creative thinking among students, stimulating a reflective and flow state during the assignments and improving communication and egalitarian participation among remote peers.

The students developed new ways of tackling challenges and entrepreneurship problems by building metaphors and analogies, demonstrating the creative power of tactile engagement in remote working. Such student behaviour during lockdown has been also reported by Karaca-Salgamcioglu and Genç (2021) and has been described as hacking the studio, referring to when students used any means available to them at home to compensate for the lack of access to other materials, tools, and laboratories at university.

Theoretical Framework

Business innovation methods such as Lego Serious Play (1996) and Playmobile Pro (2015) are based on constructionism. Lego Serious Play is a process that creates 3D models to communicate ideas, concepts, and thoughts to fellow participants (Stephen, 2018). On a similar note, the use of modelling clay in HEI learning environments has recently proved that it can lower the negative feelings of students towards a subject by 100%, increasing the student's understanding and enjoyment of the module (Lace-Costigan, 2017). Adults play primarily for the experience of participating, rather than for the learning outcomes they can achieve through their participation (Henricks, 2020). Relevant to this study is the so-called "flow" state. During flow, the person can sustain prolonged attention to the same goal due to the alignment of thoughts, intentions, and feelings (Csikszentmihalyi, 1990). The flow state, according to Dietrich, merges perceptions and actions, integrating the brain's sensory input with the motor output beyond total self-involvement. This eliminates distractions like worries of failure or fantasies of success (Dietrich, 2015, p. 164). Therefore, to achieve the flow state in creativity, one needs a challenging task that requires intense concentration, a skilled movement sequence, clear goals, immediate feedback, the elimination of distractions from consciousness, and the disappearance of self-consciousness and worry (Csikszentmihalyi, 1996, pp. 215–219).

Cognitive Learning Styles

Cognitive style refers to the manner or form of cognition (Muhammad et al., 2015) with importance in how people generate and implement ideas (Rebelo et al., 2021). Awla described three categories of learning styles—cognitive, personality, and sensory (2014, pp. 241–242) (Fig. 7.1).

Cognitive learning style encompasses attributes such as analytical/global, field dependent/independent, impulsive/reflective types. Personality learning style encompasses attributes like extroverted/introverted, random-intuitive/concrete, sequential, and closure-oriented/open-oriented types. The sensory learning is divided into visual, tactile/kinaesthetic, and auditory learning sub-types (Dornyei, 2005).

The Figurative Communication in the Classroom

Figurative language is developed to appeal to the imagination in order to convey a complex meaning with little reference to literal meaning since

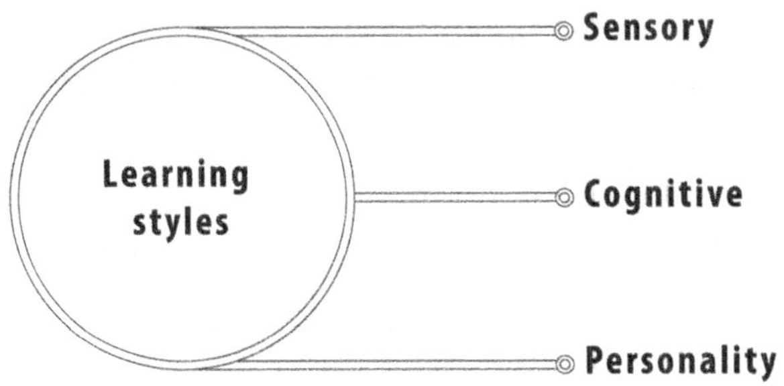

Fig. 7.1 The three categories of learning styles—cognitive, personality, and sensory (based on Awla, 2014)

it refers to the imagination (Hutauruk, 2019). Metaphors are a form of figurative language that convey multiple layers of meaning (Ward et al., 1997). In addition, they involve a comparison of thoughts that invites the receiver to approach something from a different point of view (Steen, 2007). On the contrary, literal communication is denotative such that the meaning can be found in a dictionary and does not involve any kind of interpretation. Metaphors communicate complex topics effectively, helping people to understand and vividly visualise unfamiliar concepts (Reference Yourdictionnary.com). According to Ward et al. (1997), the underlying process of creative thinking comprises conceptual combination, metaphors, analogy, and mental modelling. The importance of metaphors in creativity is great (Brand, 2013), giving individuals more tools to break, blend, or bend concepts from different disciplines and backgrounds.

Multi-dimensional Metaphors—Enhanced Communication

The Cognitive Metaphor theory argued that humans not only communicate using visual or auditory metaphors, but also, use metaphors to think. Metaphors are present in other forms of communication such as pictures, music, sounds, gestures, and touch (Forceville, 2006). Multi-dimensional metaphors are used in our everyday life. We use sound, music, pictorial, spoken and written signs, gestures, sounds, music, smells, tastes, and touch to communicate a comparison between elements or to facilitate the understanding of an idea. Touch metaphors can evoke the idea of proximity and feelings, such as "get in touch" or "she's got a touching story to tell". The touch metaphors are so common in the English language that people are hardly aware of them. We take touch for granted.

Imagine the following scene. John brings his new mobile phone to class and starts showing it off to his peers. Fiona is curious, she wants to take a closer look, so she asks, "Can I see it?" while extending her arm to reach out to the mobile phone. What she implied was "Can I touch it, feel it, and make sense of the features and characteristics while I feel the overall construction, texture, weight, and materials of the mobile phone?".

Research Methods

This study utilised two approaches: the social constructivism and the social constructionism. Social constructivism posits that knowledge is constructed through participants' active interaction (Schreiber & Valle, 2013). To create knowledge, according to Vygotsky (1978), individuals share a framework comprised of language and culture that help them understand reality and define sense. In other words, knowledge is created by interacting with others who share language and cultural information. On the other hand, social constructionism involves the idea of learning by making. In other words, social constructivism focuses on the personal learning that occurs through group interaction, whereas social constructionism focuses on the artefact that is created through group interaction. In the social constructivist theory, shared meaning among individuals creates the knowledge of reality we have. Our identity is created by the interaction we have with the environment and society and the reaction of the expectation of society.

Designing the Activities

As the first step in designing online activities for this module, we compared the missing elements from the teaching scenario before Covid-19 with those from the initial response to Covid-19 restrictions. The categories compared were activities for student engagement, cognitive styles, and teaching techniques. This comparison made visible the loss in student engagement resulting from the remote student and staff interactions and made clear that students with different learning styles were left out in the initial scenario. The study was conducted in a module called Design, Innovation, and Entrepreneurship, delivered to second-year students from multiple disciplines. This module has been created following the suggestions of Valencia and Pearce (2019) and Valencia et al. (2021) on Design Entrepreneurship and the Designer's entrepreneurial journey, making empathy, context, and problem understanding essential for entrepreneurship. The image below shows the regular teaching scenario under normal conditions for the module before the Covid-19 pandemic.

As Fig. 7.2 shows, the face-to-face module delivery favoured many activities that promoted student engagement, making room for all the different learning styles and teaching techniques. In contrast (Fig. 7.3), the initial response to the pandemic was reactive and caught many lecturers off guard. The availability of teaching tools and the staff's familiarity with new online classes significantly affected the teaching techniques. The online-teaching scenario for the module decreased the number of activities that promote student engagement, excluding a good number of students with learning styles other than verbal, visual, and auditory. In addition, teaching techniques were severely reduced in class delivery.

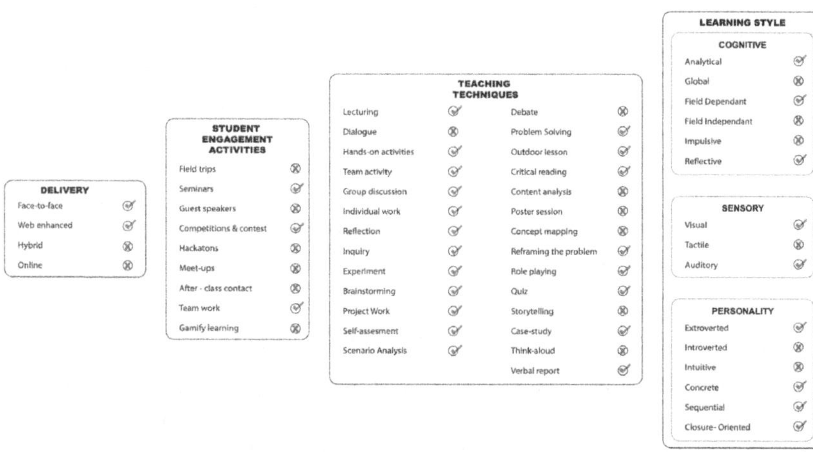

Fig. 7.2 Diagram of the normal teaching scenario before the Covid-19 pandemic for the chosen module

Fig. 7.3 Initial online teaching scenario under Covid-19 restrictions

Re-envisioning the Remote Classroom

To envision a new type of online module, we gathered inspiration from the corporate visions of future tech depicted by corporate giants such as Huawei and Microsoft. According to these visions of the future, people's lives are dominated by screens, displays, and "click-points" rather than touch points. In their visions, people are drowned in data that is continuously displayed at the kitchen table, in the lift, on office walls, and important decisions are taken by the second. Having said that, it is reasonable to say that running only online modules for the last two years has been daunting for students and faculty since the pandemic started. How do you feel every time you see that your calendar is full of online meetings, one after the other? Let alone, how you feel trying to engage with students without seeing their faces, being able to see only their acronyms, and listening to random "yes", "nope", "is my mic on?", and "see you"?

What If? and How Might We?

According to the Design Council (2020), the three H model can describe the Design discipline. The head refers to the problem-solving ability of it, which taps into the visualisation and conceptualisation of intangible elements; the heart encourages placing people at the centre, building

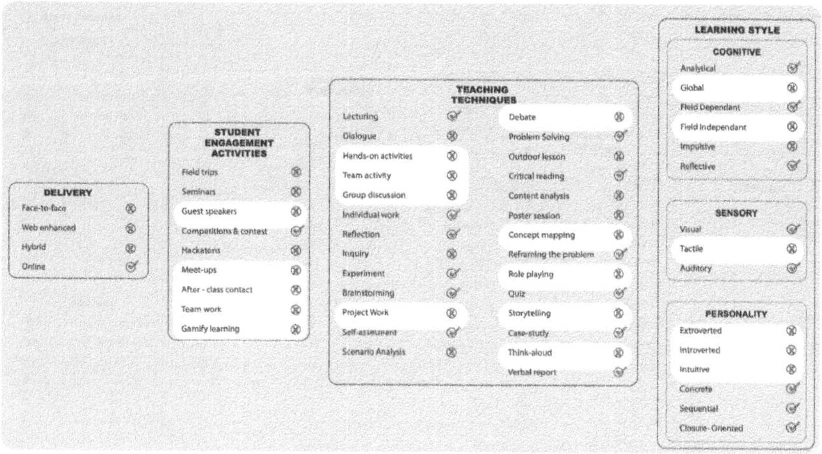

Fig. 7.4 Identified opportunity areas to improve the Student Experience in the module

empathetic solutions collaboratively; and finally, hands symbolise thinking through making, turning the invisible into visible and the complex into understandable through the power of making. At Mi:Lab, we discussed the possibility of going against the flow, tapping into the possibilities brought up by virtual integration. Our quest led us to this question—What if we could bring back the tactile experience to the online classroom (Fig. 7.4)?

After brainstorming multiple possibilities, we produced the following questions to help us navigate this challenge. How might we bring the other five learning styles into the online classroom? How might we embed the tactile experience into the visual-verbal class delivery? We had a hunch that maybe the tactile metaphors could encourage other cognitive learning styles in the online learning experience.

The Activities

In Higher Education, haptic elements are rarely used outside of arts and design. The concept of thinking through making puts tactile learning back into the equation. The goal of these activities was to improve student engagement by including competitions and contests, after-class contact with the lecturer and peers, and teamwork; increase the number of learning styles that were missed out in the initial phase of the pandemic: the Kinaesthetic and Touch, Intuitive, Sensing and Group learning styles; and use more teaching activities like hands-on and team activities, brainstorming, concept mapping, and think-aloud protocols. To address the requirements proposed before, four major activities (Table 7.1) were designed for this module.

The planned activities focused on the emergence of metaphors to spark creative thinking among members of each team. Furthermore, the activities focused on developing the business model, exploring the value proposition, identifying the user, the problem, and the solution, as well as cost structure and revenue streams. The team members built these business model sections remotely while they had a virtual session on Microsoft Teams. Due to the strict restrictions imposed in Ireland and to minimise the interaction of our students, we put together all the required materials to be delivered to each student (Fig. 7.5).

Students were sent a parcel to their home address containing a Lego Serious Play kit, a kitchen timer, Play-Doh, and the instructions to follow the class activities.

The Delivery of the Activities

The activity took place in semester one in 2021. The activities were delivered in four sessions. All the students enrolled in the module participated; 32 Students in 9 teams were connected during the sessions. Each team was assigned a breakout room in Microsoft Teams during each session to discuss the activities. Each session was recorded to allow students to get the content asynchronously and use it as a support for their individual and teamwork. The mood boards and research insights were also shared in a digital board using the web-based app called Miro.

The students had to work from home while building a collective understanding of their Business Idea using the items sent in the parcel. Students used the Play-Doh and Lego bricks to share their understanding of their ideas and contexts. Moreover, the rapid prototyping happened with the use of the Play-Doh and Lego bricks. The researcher organised 20-minute online feedback sessions with the students to ask about their engagement, teamwork, attention span, and creative flow.

Table 7.1 Proposed multimodal learning activities for the delivery of the module

Activity	*Aim*	*Description*	*Student engagement activities*	*Learning style*	*Teaching techniques*
Activity 1 Individual—3 hrs **Warm-up activity—Three-dimensional context**	Develop building skills and break the ice among the class	Each student must create a Play-doh model of the following themes and then share their ideas with the rest of the group through an integrative story What would you do with that money if you won the lottery? • What does "Young Adult" mean to you? • What does "Covid-19" mean to you? • What does "Online Learning" mean to you?	Gamify learning	Touch/Kinaesthetic, intuitive, active	Hands-on activity, Think-aloud

Activity	*Aim*	*Description*	*Student engagement activities*	*Learning style*	*Teaching techniques*
Activity 2 Team 4 hrs **Value proposition**	Build a three-dimensional Value Proposition canvas with the help of your team	Each team must work together to build each section of the Value Proposition canvas using Lego Bricks. After building each section of the VPC, the team will connect all the elements into an integrative story. Elements of the VPC are: • Jobs to be done: Functional, Social, and Emotional • Gains • Pains • Gain Creators • Pain Relievers • Product and Service propositions	After-class contact, teamwork	Touch/Kinaesthetic, intuitive, active, group	Hands-on, team activity, brainstorming, activity, concept mapping, Think-aloud

(continued)

Table 7.1 (continued)

Activity	*Aim*	*Description*	*Student engagement activities*	*Learning style*	*Teaching techniques*
Activity 3: Team 4 hrs **Creation of New Scenarios**	Build the rest of the elements of the Business Model canvas using Lego bricks	Each team must work together to build each section of the Business Model canvas. Prototype at least three variations of the Business Model canvas. Connect all the elements into an integrative story. Elements of the Business Model canvas: • Key Partners, Activities, and Resources • Customer Relationships • Channels • Cost and Revenue Streams	After-class contact, teamwork	Touch/Kinaesthetic, intuitive, active, group	Hands-on, team activity, brainstorming, activity, concept mapping, Think-aloud

Activity	*Aim*	*Description*	*Student engagement activities*	*Learning style*	*Teaching techniques*
Activity 4: Group 4 hrs **Imperfect Variations**	Once you and the team have created a Business Model canvas, it's time to imagine imperfect variations of it	Each team must work together to imagine five different variations of the Business Model canvas. Elements to include in the variations What elements would change if • Disney was creating this business? • Uber was creating this business? • Nintendo was creating this business? • Armani was creating this business? • Apple was building this business?	After-class contact, teamwork	Touch/Kinaesthetic, intuitive, active, group	Hands-on, team activity, brainstorming, activity, concept mapping, Think-aloud,

Fig. 7.5 The students received physical and digital kits with prompts

Analysis

We used thematic analysis to analyse the data. Thematic analysis is compatible with the research paradigm selected for this study. As Braun and Clarke (2006) proposed, the thematic analysis offers a theoretically flexible approach to analysing qualitative data and should be seen as a foundational method for qualitative research (p. 4). As thematic analysis suggests, we concentrated on interesting data about the tactile activities and students' experiences during the sessions to identify themes and patterns that were important to the study (Maguire & Delahunt, 2017).

Given the experimental nature of the study, we concentrated on the semantic analysis of the experience rather than the latent meaning of the gathered data. Table 7.2 shows the resulting table of themes and codes from the analysis phase.

The students reported the need for more time at the beginning of the workshop to level up the brick building skills in each team. Having no experience building or playing with the given materials reflected in the creative confidence reported by students at the onset of activities. Some students expressed their lack of creative confidence at the outset of the activities. The build-aloud protocol—A combination of the

Table 7.2 Themes and codes

Theme	*Code*	*ID*	*Verbatim data extract*
Need for pre-skill-building session	Amateur making	8-BM	Not everyone has played with Legos or play-do [Play-doh] before
	Practising voice-hand articulation	5-TL	Doing this [building the model while thinking aloud] is harder than I thought. It took me some time to get my head around this way of sharing ideas, but it is natural once you get it. It is like cooking and chatting with your friends at home
Effortless achievement	Missing interaction	4-CB	[About the tactile activities] It was fun, engaging, and hands-on; something I have missed during online classes
	Satisfying activity	5-TL	It was fun, engaging, and empowering
Sense-making	Hands-on explaining	8-VC	Throughout the workshop, it was encouraged to speak through the model, instead of using words whenever we were trying to address a problem or explain something
Attention inhibitors	Digital fidgeting	6-CM	Usually, I struggle in online classes. I'm easily distracted by my phone; this approach is spot-on
	Physical fidgeting	9-RA	I have something to fiddle with that keeps me away from biting my nails or checking my phone

(continued)

Table 7.2 (continued)

Theme	*Code*	*ID*	*Verbatim data extract*
Other uses	Using this approach as a platform	1-AW	I wish I could have more modules exploring new methods like this one right now

think-aloud protocol (Ericsson & Simon, 1980) while building a model—presented a heavy cognitive load in some individuals. 73% of students noted the benefit of using three-dimensional models during online meetings. To fully tap into the power of tactile activities, students had to be reminded to use their models to build up their arguments, explanations, and descriptions. One side benefit of the time-lapse video was that the student did not want to interrupt the footage during the building times, giving at least a 3-hour window without using their phones during the sessions.

Results

This project provided solid evidence that tactile learning can positively affect the interaction between students while working remotely. A final probe among students yielded the following results: Project information is gathered from three sources. The time-lapse video played a key role in improving concentration during the activities, since the students had to get away from their mobile phones for the entire session (3–4 hrs) and while they were with their peers. Figure 7.6 shows snaps of two video submissions.

The Lego model allowed the team to build collective landscapes of problems and concepts remotely (Fig. 7.7).

The models helped the students visualise and concentrate on the big picture, the contexts, and the stakeholders of their project. Taking online classes asynchronously allows students to learn at their own pace, rewind the video until you get the concept right, and learn at their convenience. However, the asynchronous modality cripples the team's work and interaction, due to the timetable mismatch and absence of common ground.

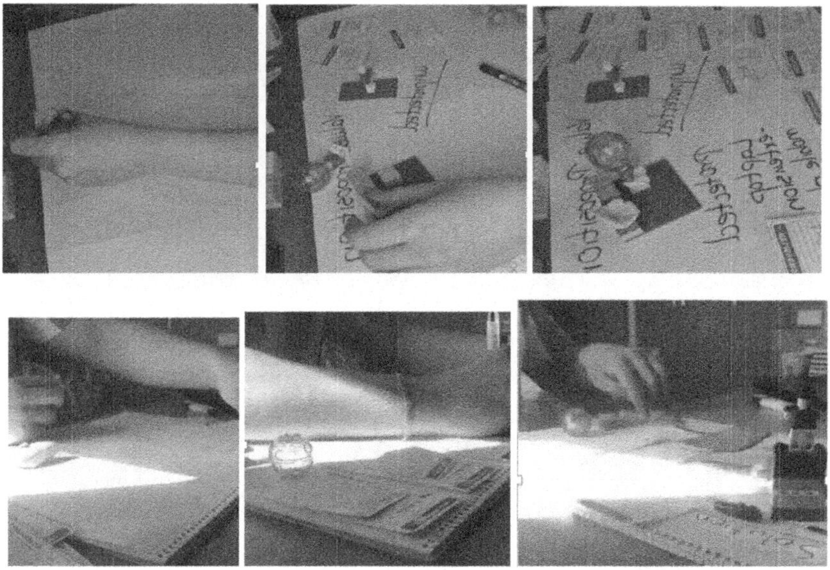

Fig. 7.6 Two sequences of the time-lapse video submissions

Conclusions and Practical Implications

Online learning did not sit well with many of the students who wanted to have a closer connection with their peers and lecturers. To alleviate this, we suggest that other learning styles could be considered when designing online content and delivery at any level of the education system. Tactile online learning can also improve the range of activities that lecturers can explore in their modules, helping students with different personalities engage in discussions and enabling other types of communication.

Evidence from this study proved several benefits of the tactile online learning approach over the reactive learning approach that resulted from the quick migration of face-to-face classes to online learning during the Covid-19 pandemic. Utilising a tactile approach to online learning can assist students in the sense-making of new concepts. However, we believe that the principal value comes from the visual narratives shared among the participants. This approach proved that it fostered the confidence to speak up in diverse audiences. Also, storytelling can be enhanced by the tangible metaphors built in the ideation phase of the business modelling. The data

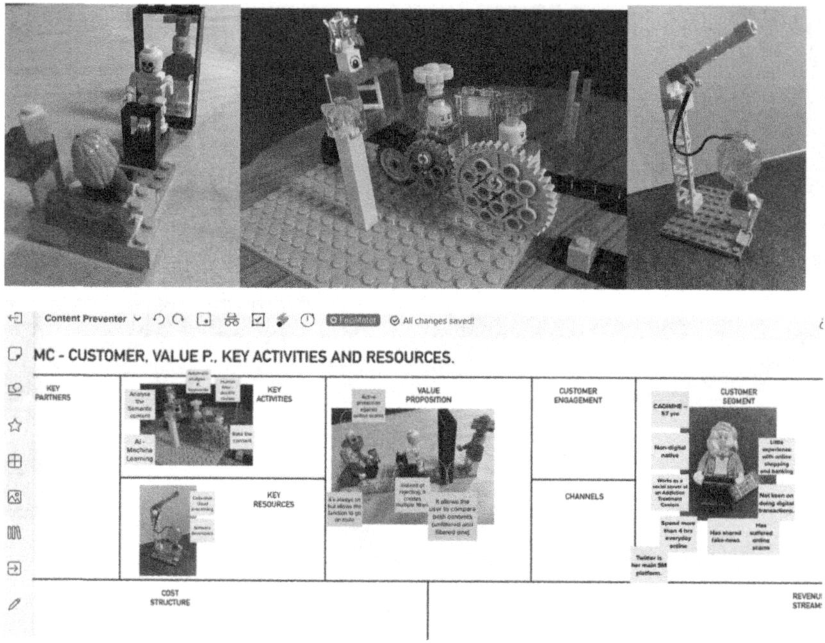

Fig. 7.7 Models built by different students at home: Value proposition model (top-left); Key activities (top-centre); Key Resources (top-right). Models over the Business Model Canvas (Osterwalder & Pigneur, 2010) on an online collaborative whiteboard (Bottom)

indicates that by using tangible materials such as bricks and clay, students could materialise their ideas, promoting a collective understanding among their team members.

References

Awla, H. A. (2014). Learning styles and their relation to teaching styles. *International Journal of Language and Linguistics, 2*(3), 214–245.

Brand, J. (2013). Cognition of creativity. In E. Carayannis (Ed.), *Encyclopedia of creativity, invention, innovation, and entrepreneurship* (p. 230). Springer.

Braun, V., & Clarke, V. (2006). Using thematic analysis in psychology. *Qualitative Research in Psychology, 3*(2), 77–101.

Csikszentmihalyi, M. (1990). *Flow: The psychology of optimal experience*. Harper & Row.

Csikszentmihalyi, M. (1996). *Creativity: The psychology of discovery and invention*. Harper Perennial.

Design Council. (2020). Using design as a force for change. Accessed on Accessed on 5th Feb 2023. https://www.designcouncil.org.uk/fileadmin/uploads/dc/Documents/Design%2520Council%25202020-24%2520Strategy.pdf

Dietrich, A. (2015). *How creativity happens in the brain*. Palgrave Macmillan.

Dornyei, Z. (2005). *The psychology of the language learner: Individual differences in second language acquisition*. Lawrence Erlbaum Associates.

Ericsson, K. A., & Simon, H. A. (1980). Verbal reports as data. *Psychological Review, 87*(3), 215–251.

Forceville, C. (2006). Non-verbal and multimodal metaphor in a cognitivist framework: Agendas for research. In G. Kristiansen, M. Achard, R. Dirven, & F. Ruiz de Mendoza Ibàñez (Eds.), *Cognitive linguistics: Current applications and future perspectives* (pp. 379–402). Mouton de Gruyter.

Henricks, T. S. (2020). Play studies: A brief history. *American Journal of Play, 12*(2).

Hutauruk, B. S. (2019). The use of figurative languages on the students' poetry Semester V at FKIP Universitas HKBP Nommensen. *Journal of English Language and Culture Hasil Penelitian, 9*(2), 128–137.

Karaca-Salgamcioglu, B., & Genç, I. (2021). The ones who have never been physically in a studio. *Design and Technology Education: An International Journal, 26*(4), 130–143.

Lace-Costigan, G. (2017). Perceptions of play: Using Play-Doh to enhance the student experience in bioscience higher education. *International Journal of Game-Based Learning, 7*(3), 26–37.

Lehtonen, M. J., Khamisani, N., & Gatto, G. (2021). Playful absence/absence of play. *Design and Technology Education: An International Journal, 26*(4), 20–36.

Maguire, M., & Delahunt, B. (2017). Doing a thematic analysis: A practical, step-by-step guide for learning and teaching scholars. *All Ireland Teaching and Learning in Higher Education Journal, 9*(3), 3351–33514.

Muhammad, T., Sarojini, E. G., & Abdurauf, R. (2015). Cognitive styles field dependence/independence and scientific achievement of male and female students of Zamfara State College of Education Maru, Nigeria. *Journal of Education and Practice, 6*(10), 58–63.

Osterwalder, A., & Pigneur, Y. (2010). *Business model generation: A handbook for visionaries, game changers, and challengers*. Wiley.

Rebelo, S., Marques, C. S., & Santos, G. (2021). The influence of cognitive styles as promoters of entrepreneurial orientation and intrapreneurship as drivers of innovation: The case of pharmacy professionals in Portugal. *Sustainability Journal, 14*(1), 368.

Schreiber, L. M., & Valle B. E. (2013). Social constructivist teaching strategies in the small group classroom. *Small Group Research*, *44*(4), 395–411.

Schwab, K. (2016). The Fourth Industrial Revolution: What it means, how to respond, World Economic Forum. Accessed on 5th Feb 2023. https://www.weforum.org/agenda/2016/01/the-fourth-industrial-revolution-what-it-means-and-how-to-respond/

Steen, G. J. (2007). *Finding metaphor in grammar and usage: A methodological analysis of theory and research.* John Benjamin's Publishing Company.

Valencia, A., Lievesley, M., & Vaugh, T. (2021). Four Mindsets of Designer-Entrepreneurs. *Design Journal, 24*(5), 705–726.

Valencia, A., & Pearce, A. (2019). Notes on designer entrepreneurs and The Geppetto effect. *Design Management Review, 30*(2), 16–21.

Vygotsky, L. (1978). *Mind in society*. Harvard University Press.

Ward, T. B., Smith, S. M., & Vaid, J. (1997). Conceptual structures and processes in creative thought. In T. B. Ward, S. M. Smith, & J. Vaid (Eds.), *Creative thought: An investigation of conceptual structures and processes* (pp. 1–27). American Psychological Association.

CHAPTER 8

Designing Learning Design Pedagogy: Proactively Integrating Work-Integrated Learning to Meet Expectations

Camille Dickson-Deane, Keith Heggart, and Robert Vanderburg

Introduction

As universities are facing significant challenges in the twenty-first century (Kaplan, 2021) preparing for the future and what it may bring would require decisions that are yet to be identified. The awareness of unknown decisions means that future skills are difficult to qualify for future work. With the rise of competing organisations, the increase of *other* credentials,

R. Vanderburg
CQ University Australia, Bundaberg, QLD, Australia

C. Dickson-Deane (✉) · K. Heggart
University of Technology, Sydney, NSW, Australia
e-mail: Camille.Dickson-deane@uts.edu.au

K. Heggart
e-mail: keith.heggart@uts.edu.au

M. J. Lehtonen et al. (eds.), *Design Education Across Disciplines*,
https://doi.org/10.1007/978-3-031-23152-0_8

and the separation between employment and tertiary degrees, universities are grappling with their position in society and the role of education as a social good (Kaplan & Kaplan, 2021a, 2021b). In the face of problems like climate change, mass inequality, and the movements of large numbers of people across the globe, the traditional disciplinary silos common in universities are struggling to adapt or offer solutions. This has led to an interest in multi-, inter-, intra-, and transdisciplinary research (Bliemel & van der Bijl-Brouwer, 2018), where there is a focus on skills that are less discipline-specific, and instead emphasise transfer, creativity, critical thinking, and resilience. A number of universities have adopted these principles into their graduate attributes (*The University of Edinburgh's Graduate Attributes*, 2021; *UTS Model of Learning*, 2014), stating that they believe these kinds of skills are essential in the development of graduates as they prepare to enter society. From this we can view the many universities that are adopting centralised models of teaching and learning, with expectations regarding the use of Learning Management Systems (LMSs) as a support to this front. This is stimulated, in part, by the findings of the Sloan Consortium Reports (Allen & Seaman, 2008, 2013), Bradley Reports (Bradley et al., 2008), multiple Horizon Reports (*2005 Horizon Report | EDUCAUSE*, 2005; *2006 Horizon Report*, 2006; EDUCAUSE (Association), 2020), which raised questions about the experiences of students and the importance placed upon teaching and learning experiences at the tertiary level. Years later, universities around the world are seeing the value in employing learning designers to assist academics and make this transition into blended or fully online modalities (Nworie, 2022). Something similar is taking place within the public sector, where training and development professionals, either within organisations or as outsourced third parties, are increasingly discussing their work in terms of learning and performance design, and especially workplace elearning design. Indeed, the field itself is expected to grow significantly over the next decade (Johnson, 2020).

These changes were occurring before the disruption induced by COVID. The pandemic has only accelerated this demand, as more and more organisations examine the validity and suitability of online and blended models of learning and development. While the initial phase was more in line with emergency remote teaching (Heggart, 2022), more carefully planned models, often requiring design expertise, are now being trialled in different contexts. Contexts are the foundation and thus have a direct dependency when one thinks of designing for learning and performance. Understanding the factors that influence how a context shapes the learning experience helps improve outcomes for institutions and individuals (de Alvarez & Dickson-Deane, 2018; Romero-Hall et al., 2020).

Designing for learning [and performance] requires an in-depth knowledge of how learning occurs (Donovan et al., 2000; Ormrod, 2004), thus building on learning sciences, cognitive sciences and psychology, education, and other nuanced disciplines. It [designing for learning] becomes an even more complex process when technologies which include those powered by electricity, hosted on networked systems (i.e., intranets, the Internet, etc.), accessed by [mobile] devices through an interface, and provisioned to learners via digital learning systems/environments are involved. Designing for learning and teaching is iterative and almost interwoven in its existence, as the process used to instruct also informs the content being learned. The difficulty in producing learning designers who are knowledgeable and skilled in the profession has always been a challenge to the field (Tripp, 1994). Understanding how learning design skills are to be designed to achieve proficiency is due to the associated knowledge being more tacitly formed than compared to knowledge that is explicit in nature (Celik, 2021; Wilson et al., 1993). By situating the knowledge to be learned within contexts, students can easily grasp concepts by creating a bridge between the learner and the contextualised content (Dickson-Deane, 2023), thus creating opportunities for transformative capacities in pedagogical designs.

Work-integrated learning, also known as cooperative education in some geographies, depends on the integration of the disciplinary and societal context to add the value needed for the learning experience (Saunders, 2019). As students matriculate into tertiary-level institutions, the institutions need to plant social good throughout the curriculum, which requires an awareness of the skills demanded and an ability to design deliberate yet strategic activities. The skills of transfer, criticality, creativity, and resilience will then permeate the curriculum through a systematic approach with the aim of creating a networked design between tertiary institutions, general society, and industry. Knowing how this may look from a design perspective and then measuring it against the outcomes that are achieved to see if they meet the needs of industry and society at large is the next step to have meaningful translation (Carr-Chellman & Carr-Chellman, 2020).

Incorporating Work-integrated Learning into Learning Design as a Strategy

Perhaps appropriately for a profession that continues to struggle to define itself (Rieber, 2018; Wagner, 2011), learning design education and training and the boundaries that encompass the field remains a confused space. While there is a profusion of different credentials and programs catering to learning design and technology expertise in countries all around the world, there still remains contention about what should be incorporated into these programs, and what kind of skill set constitutes what is required by a neophyte learning designer or technologist (Heggart & Dickson-Deane, 2021). Internationally, this has been a problem for at least three decades (Rowland, 1992; Tripp, 1994), and has stimulated a number of approaches, including problem-based methods (Dickson-Deane & Asino, 2018, p.; Silber, 2007), studio approaches (Cross, 2011), and studies of practice (York & Ertmer, 2016).

The lack of clarity has also led to some frustration on behalf of students. One common refrain (Gardner et al., 2021) is that students often feel that they are not adequately prepared for interviews, for learning designer roles, or for working as a learning designer in the field. The argument by students is that courses focus too much on theoretical considerations—the "why" of learning design, and not enough on practical matters—the "how". This means that the curriculum which focussed on specific tools and technologies that might once have been of value in a course for learning designers (and this includes both technological and procedural tools) might be quickly becoming obsolete, or replaced by alternative preferred approaches or tools—focussing on the skills towards use of a specific tool as opposed to the pedagogical value of a type of tool. The design of learning design curriculum to support the much needed transference of skills is an important requirement here. Skills that are needed not just for employment but for life-long and life-wide living learning by bridging the divide between the knowing of "why" towards the applicability of "how to do". A good solution is the use of internships or cooperative education as it is called in some locales. However, there are growing concerns about equality in terms of access to and designs of internships (Rainford, 2010), their appropriateness for a changing student demographic—a problem which is exacerbated in postgraduate settings (Chatterjee et al., 2019), and student demands for more effective and meaningful learning experiences. Post-2020, of course, there is the added challenge of providing internships in an environment that might be facing limited interaction and lockdowns due to the ongoing threat of the COVID19 pandemic. These problems—as challenging as they are—can be approached from the perspective of a learning design problem. Through reconsidering the nature of the course design and especially the elements of work-integrated learning present within, it is possible to develop a design for learning that focusses on profound and transformative learning and transferable skills and knowledge.

Modulating the *Design and Applicability* of Learning Design Curriculum

Achieving the experiential value of the skill of learning design requires an in-depth review of the curriculum and how the pedagogy actuates the curriculum. This in part requires academics in the field to fully review and then embed their core understandings of the field into their own semantic understandings for teaching. As this process is actioned through pedagogy, the mapping of key knowledge to an existing schema can be difficult to attain (Rumelhart, 1984) and developing a method through which there is meaningful comprehension can be difficult to design. In order to design such, conditioning the mind to accept the knowledge and truly comprehend it, requires that the brain has to be readied. Designing for transfer requires the understanding of how learning occurs (Ormrod, 2004) and the complexity required for the best outcomes (Jackson et al., 2019; Sangwan & Garg, 2017; Thorndike, 1924). To facilitate applicability, transfer must occur and the belief that transfer can only occur between similar context and conditions will soon be replaced with the notion of connectionism—the understanding that through sufficient practice and enough stimuli to motivate and promote learning new pathways a fertile bed suitable to accept knowledge can be created to enable transfer processes (Reed et al., 1974).

As one learns the skills embedded in the field of learning design, there are three approaches that can be used to increase the knowledge needed; (1) one can convert what is known into a map, compare the cognitive map against others, and create new knowledge (i.e., analogical transfer), (2) take the current knowledge, reduce the steps that form the foundation of said knowledge, and rework it until it is optimised with a specific context for a level of complexity (i.e., knowledge compilation), or (3) see the current knowledge when matched against another solution as a form of corrective action to be attained (i.e., error correction) (Anderson et al., 1981; Chen, 2002; Ohlsson, 1996). Through extensive dialogue with students, referencing workplace tasks and the context in which they are situated, educators can then gauge how much has been transferred and the accuracy of such (Fleming et al., 2021). Dialogue about the students' needs helps the educator scaffold how the skills they teach can be transferred to the future endeavours of the students in the workplace. Learning which tasks are most appropriate for the workplace is based on the dialogue with the students and knowledge of the workplace needs. The educator can use the knowledge gained from the dialogue with the student to show them how to transfer the skills the workforce desires, so students are more work ready. Finally, the educators need to create an environment in which students feel safe and are encouraged to ask questions. The students need to feel safe enough to attempt, and possibly fail, transfer skills for transfer to be taught effectively. This freedom to fail at transfer helps the student learn how to transfer faster and more effectively. The process of the transfer will then set the foundation for whether the data can be transferred near, (i.e., to a similar problem space) or far, to a problem space that is disparate in discipline and complexity (Jackson et al., 2019; Perkins & Salomon, 1988). As these conditions are provided for a novice learning designer, the ability for skills to grow allows for an improved and more notable designer ready for the activity in an employed mode (Phye, 2001; Phye & Sanders, 1992).

When do the Curriculum and Pedagogical Design Work for Learning Outcomes?

Research has been striving to see how academic institutions can increase the ability to transfer the skills taught in academia to the workforce. Research began looking at what kind of tasks taught in academic classes were transferable with the first realisation was that collaborative activities in academic settings (Hakel & Halpern, 2005) and authentic tasks demonstrating applied learning concepts (Burke & Hutchins, 2007) aided in the transfer process over the more theoretical concepts taught in academic classes (Veillard, 2012). Theoretical concepts were more mentally demanding and, thereby, with nothing to anchor the meaning became more difficult to transfer. In addition, activities taught in academic settings which were complex (McDonald et al., 2012), because they emulate work environments, were hard to transfer as well. These findings lead researchers to see if transfer could be improved if industry involvement was increased in academic settings. Smith et al. (2014) found that academic institutions that increased industry involvement in academic settings prepared students more for the work environment and improved transfer abilities. This finding was further elucidated when Veillard (2012) revealed that knowledge learned in the classroom was harder to transfer to the workplace compared to knowledge gained from workplace learning experiences. This finding reveals that WIL could be the solution to building the schema required to support learning for workplace employment (Jackson et al., 2019; Rumelhart, 1984; Sangwan & Garg, 2017; Sweller et al., 2019).

Embedding WIL elements into both the curriculum and pedagogical design of learning design curriculum is easily an iterative and complex process. Additionally, if the current societal and/or world contexts were to be ignored, the designs for such courses will have little to no value (Dickson-Deane, 2023). Conceptualising how a design will sufficiently embed learning design elements infused with WIL tenets to facilitate individualised and contextualised transfer is key. A design which is also responsive to conditions which are unknown and unexpected is the true treasure in the design (Khan, 2021). The fragility expressed in the socio-economics in varied geographical locations beg for designs to accommodate the impossible but expect the probable. The key is can this be achieved and if so how?

Case Study: Operationalising WIL Into Tertiary Education

The Graduate Certificate of Learning Design (GCLD) at University of Technology, Sydney is a new course, having its first intake in 2020. It caters to the increasing demands for learning designers in both higher education and corporate fields. It is offered at a graduate level, and the students are drawn from school education, higher education, corporate settings, and non-government organisations. Many students are already working in learning design, or similar roles. Other students are drawn from the broader education sector (lecturers and teachers). A significant proportion (approximately 30%), however, have no experience in education or learning design. This mix of students presents both opportunities and challenges in terms of the design of the course. The students are generally older (30 +) as one would imagine with a postgraduate course, and a majority (approximately 65%) are female.

The course is delivered entirely online in as flexible a way as possible, offering a mix of synchronous and asynchronous learning materials in small, six-week subjects (Heggart & Dickson-Deane, 2021). There is no set pathway through the course, and students are able to enrol in the eight subjects in any order. It is a program of study which has multiple elements of WIL embedded seamlessly throughout the program and each individual course. Designing for transfer of knowledge in a program of study cannot be done in one course or another, but must be strategically planned, scaffolded, and thus would require an intense collaboration between practice, theory and research—industry and academia. This case utilised significant user and market research in addition to detailed stakeholder interviews with prospective students, practising learning designers, and employers from a range of industrial sectors. In addition, at the conclusion of the first cohort, further research was undertaken via interviews with graduating students to determine the effectiveness of these interventions. The effectiveness was measured in terms of them [students] gaining employment. Thus providing an avenue for the outcomes of the program to be infused and redirected as influencing factors into the curriculum and pedagogical practices. The case outline three elements which created a framework for the learning design (see Fig. 8.1).

1. Contextualised assessments tasks
2. Industry-like discussion sessions
3. Internship experiences.

Contextualising Assessment Tasks Within (Post)Graduate Education

It is hardly new to suggest that effective learning and instructional design begin with the requirement to make the learning relevant to the student (Dickson-Deane, 2023). Indeed, Merrill (2002) places such an emphasis on this that he makes it central to his principles of learning. The case is even more important for adult learners (Rothes et al., 2017) and those who have a choice about their learning. Relevance increases motivation (Keller, 1987) for the students. However, despite this, attempts to make

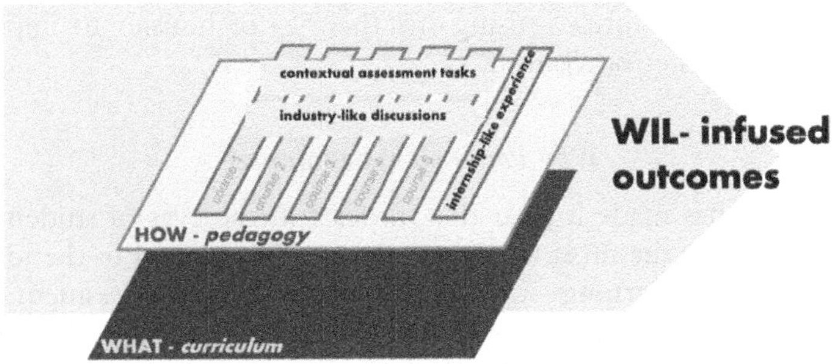

Fig. 8.1 Proposed framework for embedding WIL elements into GCLD curriculum and pedagogy

courses relevant often struggle to go further than tokenistic surveys or questionnaires at the start of any particular course. It's relatively straightforward to ask students why they are undertaking a particular course or why they are interested in their field of study—but if this is the sum of a designer's efforts to contextualise course materials and learning, then it is likely to be self-defeating—and this is especially the case if there is no further engagement on behalf of the educators with student comments.

A more effective approach, and one that was adopted in the GCLD, is to contextualise the learning at the point of the assessment to the current workplace conditions of the student. For each of the different assessment tasks, students were asked to apply the requirements of the task to current or past projects within their current or past places of employment. For example, in one of the courses in the GCLD, students are required to redesign or to develop a new assessment task. In a traditional learning design course, it would be common to provide students with a typical brief or scenario, and require them to work from that. While this may be effective, it does little to allow students to apply what they have learnt to a setting beyond the university. How much more effective is it for the students to be given the chance to apply their learning about assessment to training materials with which they are already familiar. Basically, this approach is a mechanism by which the course designers for the GCLD explicitly designed for students to apply their learning in a new context, with the idea that if they could transfer it from the university to their workplace, it would be easier for them to transfer it to other workplaces and settings in the future—using the theories of transfer to implement near and far transfer outcomes.

Why Focus on Assessments?

Biggs (1996) has made it clear that, at least in the eyes of students, it is assessment that is the most important factor. This has led to the idea that assessment drives learning—i.e. students learn for the assessment, rather than being assessed on their learning. This has led to ideas such as making contributions to discussions assessed, or assessing readings, and so on, in order to capitalise on the student focus on assessment. Debating the merits of any of these approaches is beyond the scope of this discussion as the attempt is to highlight approaches to achieving the best for the learning outcomes.

However, simply designing for this transfer is not enough. The designers of the course wanted to promote reflection amongst the students, wherein they were required to consider how effectively they had engaged in understanding the value of what they understood and believed they learned (Eraut, 1994; Schön, 1983) in order to be competent in the field. This was accomplished via short reflective pieces that were attached to each main assessment task. Thus serving as a way to embed some of the skills that are most in demand amongst learning designers such as critical thinking, design-decision making and justification, human-centred skills, etc., i.e., all skills that are likely to be needed when employed as a learning designer.

Industry-like Discussions

While no doubt effective, the contextualised assessment tasks were only part of the work-integrated learning strategy deployed throughout the GCLD the next feature that was common within each subject was the inclusion of voices from the profession—current and practising learning designers (or people in learning design adjacent roles). This was done through the offering of industry-like discussions called: In Conversation With… and Expression Sessions.

In Conversation With… were interviews with learning designers about the topics being covered in the subject at hand. For example, in a subject called Critique: Issues in Learning Design, the In Conversation With… videos documented a learning designer discussing how designs were made to be accessible, including considering diversity, inclusivity, and more prosaic topics like contrast and transcripts. These interviews were segmented and spread throughout the learning management system which housed the course, with the effect that students almost had a more senior colleague discussing the topics with them as they made their way through the course—somewhat a form of cognitive apprenticeship. Of course, this was a somewhat passive affair; students couldn't interact with the videos in any meaningful way beyond viewing them.

However, they did provide a valuable lead-in to the second element: *Expression Sessions*. These were sessions that took place at the end of every subject. In these sessions, a practising learning designer was invited to take students through something that they had been working on, related to the subject. For example, in the subject Crunch: Learning Analytics for Performance Improvement, there was a presentation and then a workshop on how to use automated feedback to personalise student feedback. These served as small-scale cognitive apprenticeships (Brandt et al., 1993) or alternatively as design studios (Smith & Boling, 2009). Students were able to observe a learning designer at work on a current and relevant problem to their own experience and to ask questions and try to do the work themselves, drawing on their own experience. Like the contextualised assessment tasks, this was an exercise in transfer, but it was a step further than in that example; here, students were asked to consider not their own context, but the context in which the guest learning designer was working.

Internship Experiences

Internships are often cited as an example of work-integrated learning and they can certainly fulfil that function, although it should be noted that they work best when they are incorporated with other, sometimes overlooked, elements of work-integrated learning (Jackson, 2018). They [internships] also called "co-ops" are typically described as a partnership between the academic institution and industry to place students who meet a qualifying criteria into industry to gain specific work-related skills (Kagan, 1985). Even before the pandemic made face to face attendance difficult, internships had some concerns; principally, they often entrench inequality amongst students, as it is unlikely that all students will be able to make the commitment to undertake an internship. This problem is only exacerbated in postgraduate study, where most students have other responsibilities, including work and family obligations with the added notion that internships can take the form of reduced or free labour thus being interpreted as advantageous only to industry participants.

Within the GCLD, the course designers wanted to find something that was a better fit for the time-bound modern student. This needed to be something that could work around a student's commitments. The result was one subject devoted to a ***flexible internship-like experience***. This subject (Work: Learning Design Project) is the final course within the GCLD, and requires students to undertake a 20–30 hour internship-like experience. However, this experience is undertaken as part of a learning design team within the university and is based on the notion of completing a project, rather than simply attending and shadowing a more senior worker.

This has a number of advantages; firstly, the flexibility of a project approach means that students can work around their work and family schedules in order to complete the work. This makes the design more individualistically socially just (Heggart et al., 2020). In addition, by deliberately designing so that there is a project to be completed, the learning is more focussed, more targeted, and thus more likely to be of value to the student, the industry–partner, and the academic institution.

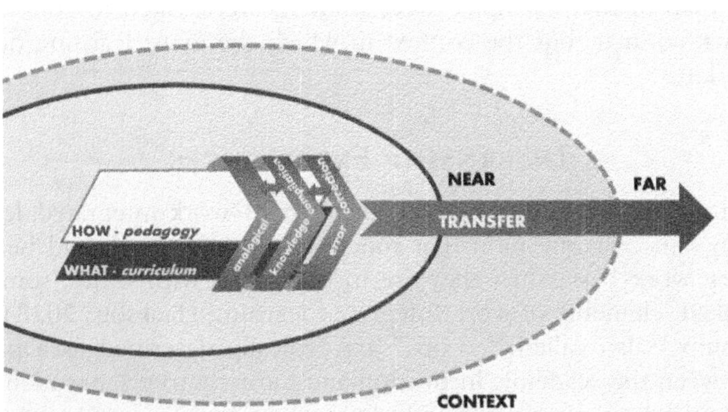

Fig. 8.2 Testing model for transferability of WIL design elements infused into curriculum and pedagogical design

Future Directions and Discussion

Given this testing model (see Fig. 8.2) and the areas where WIL elements are infused, separating the contributions towards curriculum design and contributions to the pedagogical design is imperative to measure success. WIL programs tend to measure success through employability numbers which signals to educators that their students have the qualities that current and future employers desire in their employees. As transfer is key from the educational environment to the workplace environment, there is the need to also review how the workplace environment contributes or provides feedback into the same curriculum and pedagogical designs for the program. Looking for a seamless movement in applicability in the attainment and proactive use of skill requires an entire program design that promotes an immersive experience (Csikszentmihalyi et al., 2018). A good suggestion here would be to ensure that educators are explicitly aware of what transfer designs look like and how they should be designed into the curriculum and then reinforced through pedagogical delivery (Barnett & Ceci, 2002) to achieve the expected outcome (Fleming et al., 2021). With the understanding of the type of transfer (i.e., near and far) and then the methods under which they should be measured design and delivery skills will improve even through the activity of scaffolding for transfer—educators should practise designing transfer using transfer principles (i.e., rehearsal, building meaningful understanding towards reduce error making, etc.) (Bransford & Schwartz, 1999).

As WIL is considered a learning design strategy focussing on the value of all experiences, good, bad, and indifferent is important to creating a well-rounded contributor to the industry (Kapur, 2016). There are also discussions that should be had with industries as to how to instil those desired attributes into students who are novices to the world of work, thus reframing how internships are designed and where they are strategically embedded in courses and programs for improved effectiveness. As the embedding of WIL elements is discussed, there is the opportunity to rethink how pedagogy can benefit from explicitly creating curriculum and instruction seamlessly with work-integrated learning assets—but, many can say this already occurs in some form with the likes of Google and Amazon plan to disrupt college degrees by providing job-ready on-demand learning opportunities. Maybe this model is a template almost like a powered-up just-in-time learning opportunity and with this all it means is that both university and industry need to meet at the students study desk to determine how to move forward.

References

2005 Horizon Report | EDUCAUSE. (2005). Retrieved 16 February 2022, from https://library.educause.edu/resources/2005/1/2005-horizon-report

2006 Horizon Report. (2006). Retrieved 16 February 2022, from https://library.educause.edu/resources/2006/1/2006-horizon-report

Allen, I. E., & Seaman, J. (2008). *Staying the course: Online education in the United States, 2008*. Sloan Consortium. http://sloanconsortium.org/publications/survey/pdf/staying_the_course.pdf

Allen, I. E., & Seaman, J. (2013). *Changing course: Ten years of tracking online education in the United States*. Sloan Consortium. The Babson Survey Research Group. http://sloanconsortium.org/publications/survey/changing_course_2012

Anderson, J. R., Greeno, J. G., Kline, P. J., & Neves, D. M. (1981). Acquisition of problem-solving. *Cognitive Skills and Their Acquisition*, 191.

Barnett, S. M., & Ceci, S. J. (2002). When and where do we apply what we learn?: A taxonomy for far transfer. *Psychological Bulletin, 128*(4), 612.

Biggs, J. (1996). Enhancing teaching through constructive alignment. *Higher Education, 32*(3), 347–364.

Bliemel, M., & van der Bijl-Brouwer, M. (2018). Transdisciplinary innovation. *Technology Innovation Management Review*, *8*(8), 3.

Bradley, D., Noonan, P., Nugent, H., & Scales, B. (2008). *Review of Australian higher education (Bradley report)*. AGPS.

Brandt, B. L. G., Farmer Jr, J. A., & Buckmaster, A. (1993). Cognitive apprenticeship approach to helping adults learn. *New Directions for Adult and Continuing Education, 1993*(59), 69–78.

Bransford, J. D., & Schwartz, D. L. (1999). *Rethinking transfer: A simple proposal with multiple implications* (Vol. 24). American Educational Research Association.

Burke, L. A., & Hutchins, H. M. (2007). Training transfer: An integrative literature review. *Human Resource Development Review*, *6*(3), 263–296.

Carr-Chellman, D. J., & Carr-Chellman, A. (2020). Integrating systems: The history of systems from von Bertalanffy to profound learning. *TechTrends, 64*(5), 704–709. https://doi.org/10.1007/s11528-020-00540-1

Celik, P. M. (2021). *Examination of the applied instructional design process in transforming tacit knowledge into explicit knowledge through communities of practice* [Ph.D. Thesis]. Iowa State University.

Chatterjee, D., Ford, J. K., Rojewski, J., & Watts, S. W. (2019). Exploring the impact of formal internships on biomedical graduate and postgraduate careers: An interview study. *CBE—Life Sciences Education, 18*(2), ar20.

Chen, Z. (2002). Analogical problem solving: A hierarchical analysis of procedural similarity. *Journal of Experimental Psychology: Learning, Memory, and Cognition, 28*(1), 81.

Cross, N. (2011). *Design thinking: Understanding how designers think and work.* Berg.

Csikszentmihalyi, M., Montijo, M. N., & Mouton, A. R. (2018). *Flow theory: Optimizing elite performance in the creative realm.*

de Alvarez, M. S., & Dickson-Deane, C. (2018). Avoiding educational technology pitfalls for inclusion and equity. *TechTrends, 62*(4), 1–9.

Dickson-Deane, C. (2023). Learning within context. In *Inclusive learning design in higher education: A practical guide to creating equitable learning experiences*. Routledge.

Dickson-Deane, C., & Asino, T. I. (2018, March 27). *Don't forget, instructional design is about problem solving*. EUCAUSE Review. https://er.educause.edu/blogs/2018/3/dont-forget-instructional-design-is-about-problem-solving

Donovan, M. S., Bransford, J. D., & Pellegrino, J. W. (2000). *How people learn: Brain, mind, experience, and school (expanded).* National Academy Press.

EDUCAUSE (Association). (2020). *2020 EDUCAUSE Horizon report: Teaching and learning edition*. EDUCAUSE. https://library.educause.edu/-/media/files/library/2020/3/2020_horizon_report_pdf.pdf?la=en&hash=08A92C17998E8113BCB15DCA7BA1F467F303BA80

Eraut, M. (1994). *Developing professional knowledge and competence*. Falmer.

Fleming, J., Rowe, A. D., & Jackson, D. (2021). Employers as educators: The role of work placement supervisors in facilitating the transfer of skills and knowledge. *Journal of Education and Work, 34*(5–6), 705–721.

Gardner, J. L., Snyder, D., Guilkey, J., Abbott, V., & Barclay, M. (2021). What graducate school didn't teach you about instructional design consulting. *Performance Improvement, 60*(9–10), 14–22.

Hakel, M., & Halpern, D. F. (2005). How far can transfer go? Making transfer happen across physical, temporal, and conceptual space. *Transfer of Learning from a Modern Multidisciplinary Perspective*, 357–370.

Heggart, K. (2022). A learning design model for supporting student learning needs. In *Global perspectives on educational innovations for emergency situations*. Springer.

Heggart, K., & Dickson-Deane, C. (2021). What should learning designers learn? *Journal of Computing in Higher Education*. https://doi.org/10.1007/s12528-021-09286-y

Heggart, K., Dickson-Deane, C., & Novak, K. (2020, September 7). The path towards a socially just learning design. *SRHE News*. https://srheblog.com/2020/09/07/the-path-towards-a-socially-just-learning-design/

Jackson, D. (2018). Developing graduate career readiness in Australia: Shifting from extra-curricular internships to work-integrated learning. *International Journal of Work-Integrated Learning, 19*(1), 23–35.

Jackson, D., Fleming, J., & Rowe, A. (2019). Enabling the transfer of skills and knowledge across classroom and work contexts. *Vocations and Learning, 12*(3), 459–478.

Johnson, M. (2020, May 8). Need for rapid virtual training boosts demand for instructional designers—Tampa Bay Business Journal. *Tampa Bay Business Journal*. https://www.bizjournals.com/tampabay/news/2020/05/08/need-for-rapid-virtual-training-boosts-demand-for.html

Kaplan, A. (2021). *Higher education at the crossroads of disruption: The university of the 21st century*. Emerald Group Publishing.

Kaplan, A., & Kaplan, A. (2021a). Degrees converting into lifelong learning. In *Higher education at the crossroads of disruption* (pp. 61–85). Emerald Publishing Limited. https://doi.org/10.1108/978-1-80071-501-120211005

Kaplan, A., & Kaplan, A. (2021b). From knowledge acquisition to (societal) skills development. In *Higher education at the crossroads of disruption* (pp. 37–60). Emerald Publishing Limited. https://doi.org/10.1108/978-1-80071-501-120211004

Kapur, M. (2016). Examining productive failure, productive success, unproductive failure, and unproductive success in learning. *Educational Psychologist, 51*(2), 289–299.

Keller, J. M. (1987). Development and use of the ARCS model of instructional design. *Journal of Instructional Development, 10*(3), 2–10. https://doi.org/10.1007/BF02905780

Khan, F. (2021, January 6). *#BizTrends2021: Leadership during Covid-19 and beyond*. Bizcommunity.Com. http://www.proquest.com/docview/2475565953?parentSessionId=evcfoIURmV7r0fRuaPwvVHqmDlXeEjcnTREDm%2Big0W0%3D&pq-origsite=primo&accountid=17095

McDonald, L., Leberman, S., & Doyle, S. (2012). *The transfer of learning: Participants' perspectives of adult education and training*. Gower Publishing, Ltd.

Merrill, M. D. (2002). First principles of instruction. *Educational Technology Research and Development, 50*(3), 43–59.

Nworie, J. (2022). The increasing quest for instructional designers and technologists in higher education and corporate settings. *Contemporary Educational Technology, 14*(1), ep345.

Ohlsson, S. (1996). Learning from performance errors. *Psychological Review, 103*(2), 241.

Ormrod, J. E. (2004). *Human learning* (4th ed.). Prentice-Hall.

Perkins, D. N., & Salomon, G. (1988). Teaching for transfer. (Applying knowledge to different situations). *Educational Leadership, 46*(1), 22–32.

Phye, G. D. (2001). Problem-solving instruction and problem-solving transfer: The correspondence issue. *Journal of Educational Psychology, 93*(3), 571.

Phye, G. D., & Sanders, C. E. (1992). Accessing strategic knowledge: Individual differences in procedural and strategy transfer. *Contemporary Educational Psychology, 17*(3), 211–223.

Rainford, P. (2010, November 9). The UK cannot afford to accept the proliferation of unpaid internships that cripple social mobility, entrench inequality and jeopardise our economic future. *British Politics and Policy at LSE.* http://eprints.lse.ac.uk/41570/

Reed, S. K., Ernst, G. W., & Banerji, R. (1974). The role of analogy in transfer between similar problem states. *Cognitive Psychology, 6*(3), 436–450.

Rieber, L. (2018). The proper way to become an instructional technologist. *Foundations of Learning and Instructional Design Technology.*

Romero-Hall, E., Correia, A. P., Branch, R. M. (Rob), Cevik, Y. D., Dickson-Deane, C., Chen, B., Liu, J. C., Tang, H., Vasconcelos, L., Pallitt, N., & Thankachan, B. (2020). Futurama: Learning design and technology research methods. In *Research methods in learning design and technology* (pp. 206–226). Routledge. https://doi.org/10.4324/9780429260919-14

Rothes, A., Lemos, M. S., & Gonçalves, T. (2017). Motivational profiles of adult learners. *Adult Education Quarterly, 67*(1), 3–29.

Rowland, G. (1992). What do instructional designers actually do? An initial investigation of expert practice. *Performance Improvement Quarterly, 5*(2), 65–86.

Rumelhart, D. E. (1984). Schemata and the cognitive system. In R. S. Wyer Jr. & T. K. Srull (Eds.), *Handbook of social cognition* (Vol. 1, pp. 161–188). Lawrence Erlbaum Associates Publishers.

Sangwan, S., & Garg, S. (2017). WIL and business graduate skill transfer to workplace. *On the Horizon.*

Saunders, M. (2019). Cooperative education–past, present, and future. *JADARA, 6*(2). https://repository.wcsu.edu/jadara/vol6/iss2/15

Schön, D. (1983). *The reflective practitioner.* Basic Book.

Silber, K. H. (2007). A principle-based model of instructional design: A new way of thinking about and teaching ID. *Educational Technology,* 5–19.

Smith, C., Ferns, S., & Russell, L. (2014). *The impact of work integrated learning on student work-readiness.*

Smith, K. M., & Boling, E. (2009). What do we make of design? Design as a concept in educational technology. *Educational Technology,* 3–17.

Sweller, J., van Merriënboer, J. J. G., & Paas, F. (2019). Cognitive architecture and instructional design: 20 years later. *Educational Psychology Review*. https://doi.org/10.1007/s10648-019-09465-5

The University of Edinburgh's Graduate Attributes. (2021). The University of Edinburgh. Retrieved 16 February 2022, from https://www.ed.ac.uk/graduate-attributes/framework

Thorndike, E. L. (1924). Mental discipline in high school studies. *Journal of Educational Psychology, 15*(1), 1.

Tripp, S. D. (1994). How should instructional designers be educated? *Performance Improvement Quarterly, 7*(3), 116–126.

UTS model of learning. (2014, July 24). University of Technology Sydney. https://www.uts.edu.au/about/graduate-school-health/about-gsh/uts-model-learning

Veillard, L. (2012). Transfer of learning as a specific case of transition between learning contexts in a French work-integrated learning programme. *Vocations and Learning, 5*(3), 251–276.

Wagner, E. (2011). Essay: In search of the secret handshakes of ID. *The Journal of Applied Instructional Design, 1*(1), 33–37.

Wilson, B. G., Jonassen, D. H., & Cole, P. (1993). Cognitive approaches to instructional design. *The ASTD Handbook of Instructional Technology, 4*, 21–21.

York, C. S., & Ertmer, P. A. (2016). Examining instructional design principles applied by experienced designers in practice. *Performance Improvement Quarterly, 29*(2), 169–192.

CHAPTER 9

Creating Eddies: A Transformation Design Project Founded on Caring

Joshua Korenblat

Introduction

How might we educate designers to become caring citizens who are "finely aware and richly responsible" (James, 1934, p. 149 in Nussbaum, 1985, p. 516)? This moral philosophy of fine awareness and rich responsibility, proposed by the novelist Henry James, can only be achieved by developing our faculties for listening well, imagining narratives, and understanding the particulars of lived experience. University of Chicago philosopher Martha Nussbaum studies how we might re-centre emotions in our collective life: rational, ethical, and political. The elements of clarity—concept, structure, and formal language—become one. In this way, Jamesian prose, with its acute attention to detailed imaginings, brings to life dusty philosophy. The same is true for design: as Willis (2006, p. 86) says, *design designs*. We make the designs; then, the designs make us. By practising ways of designing, we craft our disposition towards

J. Korenblat (✉)
State University of New York, New Paltz, NY, USA
e-mail: korenblj@newpaltz.edu

M. J. Lehtonen et al. (eds.), *Design Education Across Disciplines*,
https://doi.org/10.1007/978-3-031-23152-0_9

life: when to plan and when to improvise, our sense of care or control, and even our assumptions of how the world works. Design outcomes, meanwhile, reveal what we prioritize and value.

In the twentieth century, design education prioritized traditional disciplines, such as visual communication design, industrial design, or architecture. Yet even when this educational model stretched towards interdisciplinary, it often lacked a broad foundation in philosophy and ethics. Design often served commercial and business interests in creating products, messages, and services that fulfilled people's social, emotional, and functional needs. In the twenty-first century, however, design education has transitioned from a focus on objects and profit to design's end purpose in improving life: designing for experience, designing for service, designing for innovation, designing for transformation, and designing for sustainability (Sanders & Stappers, 2018, p. 17). Design educators have recognized that the "wicked problems" presented by the three great divides—ecological, social, and spiritual—require integrative learning.

Today, designers need to value philosophy, with its emphasis on ethics, and integrate ethical ideas into a more purposeful design education. When critiqued and examined, philosophy can help learners navigate over-information and debunk misinformation. Design students can turn to diverse philosophies to discover codes of conduct about how they might best live in the world, where empathy can be an instrument of compassion for people and the planet. Here, design educators aren't applying philosophy on top of their already deep and broad knowledge, skills, and values. Rather, philosophy is a foundation for a design education in the emerging disciplines, which aim to care for and improve the world. This philosophical and ethical base supports formal and informal learning settings where risk-taking and revision create vulnerability. Creative people need nurturing and care as they navigate the ambiguity of open-ended projects. The ethics that I describe fit within a broad framework introduced by feminist philosophers: the ethics of care. I will now call this ethical framework "Caring", with a capital C. Caring always entails emotion; according to Nussbaum (2015), emotion can serve as a dialogue between the particulars of an experience and guiding principles that we can apply more broadly. Emotions also create a bridge from an individual situation to more abstract principles. As a bridge and dialogue, emotions help people to work with love and reciprocity in the "potential space" and "transcend mere tepid play" (Nussbaum, 2015, p. 319). Designers can create these conditions for love, reciprocity, and play in the potential space, leading to Caring. Here, designers step away from frozen, tepid transitions and towards transformation, with ambitious yet achievable prototypes of the future.

In this chapter, I will illustrate Caring in design education with a case study of Eddy, a transformation design initiative at the State University of New York at New Paltz (SUNY New Paltz), a liberal arts university nestled in the scenic and historic Hudson Valley north of New York City. Founded by Andrea Frank, a professor of photography, and Michael Asbill, a professor of sculpture, Eddy began with a natural inspiration, the whirlpools in streams that run counter to the mainstream. In contrast to the pipeline model of education, which educates students for competitive careers and economic development, Eddy prototypes Caring relations for the campus and community. Through modelling, practice, dialogue, and confirmation—the four elements of Caring education, according to care ethics philosopher Nel Noddings (1991)—people who interplay with Eddy and create their own Eddies, can become more Caring: finely aware and richly responsible. Eddy also began with a foundation of sustainability ethics: Just Transition, which calls for transitioning from an ego-system, extractive economy to an ecosystem that values diverse voices and a more tender relationship between people and the rest of nature. Just Transition embraces the ancient Greek root of economy and ecology: *oikos*, house, (Scharmer & Kaufer, 2013, p. 67) and expands the dimensions of our house to encompass economy and ecology—people and planet. While transitioning to better ways is vital, today, designers also concern themselves with transformation, believing we need to accelerate change due to global warming, plastic pollution, biodiversity loss, and threats to democracy and vulnerable peoples worldwide, among many other social and ecological crises.

Ultimately, a design practice rooted in cultivating citizens who are finely aware and richly responsible can expand a community's flourishing. According to Nussbaum, eudaimonistic thought places the suffering person or persons among the important parts of the life of the person who feels the emotion. Here, a person might say, "They count for me. They are among my most important goals and projects" (Nussbaum, 2015, pp. 144–145). The thought "that the suffering person is part of one's own circle of concern" (ibid., p. 262) can be made stable through acts of vivid imagining; yet Nussbaum warns that emotional responses present at once an opportunity and a challenge. Eudaimonistic thought often begins with strong emotions, "invested with what we think is important and our conception of flourishing". These emotions correspond with people we know well, such as family and friends. They also arise more easily when we believe the suffering to be serious, we can't assign blame to the suffering person, and we can imagine the same thing happening to us. To reach the more distant other, "vivid presentations may jumpstart compassionate responses – for a time" (Nussbaum, 2015, pp. 144–145).

Each of us has the capacity to Care and to show compassion through action that can contribute to eudaimonia. Yet in each human, we encounter obstacles, rooted in an innate tendency towards narrow compassion, which Nussbaum (2015, p. 138) identified not only in humans but in other animals, from elephants to mice. She also elucidates emotions that impede compassion: fear, envy, and shame. Unlike other animals, humans are disgusted by bodily reminders of our animality and mortality. Humans project that disgust onto vulnerable groups of people. We project disgust onto others, evidenced worldwide in systems that segregate people by race or social caste. These impediments to Care include our tendency towards narrow compassion, which presents challenges to Caring for others at a distance. Even though vivid stories help create compassion for the distant other, research into organ donation rules shows that people favour the single vivid story over the needs of the many, which can misallocate resources and create harm (Nussbaum, 2015, p. 156). Given the fragility of our mutual ecological, social, and spiritual lives, it's worth grappling with these impediments. Caring designers can transform the world. Yet Caring is inefficient. They need to practice a responsible awareness, crafted and polished over time through observation, study, and joyful practice.

Background: Caring in Design Education and Practice

As an ethical foundation, Caring began not in academies but by observing historical household relationships. Care ethics evolved when feminist scholars, including Carol Gilligan, Virginia Held, and Nel Noddings, noticed that conventional ethical frameworks over-emphasized reason and marginalized emotion in critical thinking and right conduct. The philosophies they critiqued included foundational ones from the Enlightenment onward: deontology, the rules-based approach of Immanuel Kant, with its *categorical imperative,* and the utilitarianism of J. S. Mill, which sought to maximize satisfaction at a collective level. These rules-based, patriarchal approaches favoured rationality, moral independence, and universal principles. Yet they overlooked maternal ways of knowing: emotion, moral interdependence, and situated contexts. Historically, these elements of Caring relations exist in the uneven yet loving and reciprocal relationship between mothers and their children. Care ethicists devised guiding principles based on interdependence and relationships in this oft-overlooked setting of human development and sustainment. In any Caring situation, we can identify a person doing the caring and the cared-for. At times, this relationship flips. According to Nel Noddings, Caring begins with listening and receptivity, and in recognizing the best intentions in the cared-for. To recognize best intentions, we must empathize with the cared-for's reality. Care ethicists also scale this fundamental relational unit from the individual to a societal and ecological level.

Caring Education

Caring shares qualities with Aristotelian ethics, which unlike later ethical frameworks, richly accounted for emotion and experience. Unlike Plato, who trended towards the universal ideals, Aristotle dwelled in the particularities of experience and distrusted universal principles when considering right conduct. Aristotle considered this practice a cultivation of personal virtue. Yet Noddings and other care ethicists de-emphasize Caring as a personal virtue. Caring is "not properly labelled an ethic of virtue. Although it calls on people to be carers and to develop the virtues and capacities to care, it does not regard caring solely as an individual attribute. It recognizes the part played by the cared-for. It is an ethic of relation" (Noddings, 1991, p. 21). Noddings focuses on Caring Education, and describes four components: modelling, practice, dialogue, and confirmation. Modelling means showing Care, not just telling students about it. Listening is the key practice in modelling, and Noddings argues that listening is the cornerstone for a moral life.

In Caring Education, Noddings acknowledges that practice suffers in settings that emphasize hierarchy and competition based on extrinsic motivators like rewards and penalties. For confirmation, the final component of Caring Education, Noddings adapts philosopher Martin Buber's view. "Confirmation", she writes, is "an act of affirming and encouraging the best in others. When we confirm someone, we spot a better self and encourage its development. We can do this only if we know the other well enough to see what he or she is trying to become" (Noddings, 2006,

p. 113). Caring Education uses foundational philosophical principles, rooted in listening and reciprocity, to create more Care in interpersonal and public relationships.

Caring Design

Caring is a natural collaborator with design practice. Stanford University's d.school has exported design thinking, a flexible yet structured process for innovation popularized by the design firm IDEO. Though its frameworks vary, design thinking attempts to pattern the thinking and doing styles of creative, service-oriented designers. Yet Maurice Hamington, a professor of business at Portland State University, notes that the empathy phase in design thinking usually lacks an explicit ethical focus, moral and relational—even though it could easily accommodate ethics (Hamington, 2019, p. 100). Here, empathy expands the circle of concern beyond the person the designers aim to support to everyone involved in the design process and product.

Students of design thinking can study Caring at once as a type of knowledge and as applied, practical wisdom. In the context of design thinking, Hamington (2019, p. 97) defines Caring Design as "a human-centered innovation and problem-solving methodology/process as well as a moral and epistemological ideal grounded in a commitment to inquiry, empathy, and care for constituent stakeholders". Empathy for pain-points, a common place to begin a design thinking inquiry, transcends a mere knowledge of pain-points.

Beyond design thinking, Hamington cites a broad definition of Caring that has influenced designers in the built environment. Hamington (2019, p. 92) writes that "[t]he most commonly quoted characterization of care is offered by Berenice Fisher and Joan Tronto as 'a species activity that includes everything that we do to maintain, continue, and repair our world so that we can live in it as well as possible'".

Design typically affiliates with innovation, and innovation with technology. Yet most of the methods cited in Caring Design call to mind vintage instruments: people and time. However, beyond even time demands, taking on Care can be an investment of energy and resources. That is one reason, on an individual level, we might observe a person half-listening, or not listening at all, to another person speaking. The half-listening person might not be prepared for the psychic and cognitive investment, and the motivational displacement, that Caring might tax and require. Lange cites a challenge to Caring Design at a systemic level: deeper levels of Care can be comparatively inefficient to methods that also repair socioeconomic and ecological divides, and which promote sustainable outcomes. For example, Care is a more comprehensive and proactive system than maintenance and repair. In fact, Care might even engage with transformation design, rather than activities that nudge communities forward in a transitional way. This predicament harkens to the structure of eudaimonistic thought, where Care is richer and deeper for people we know well and who are near to us. Indeed, Caring Design concurs with Nussbaum's (2015, p. 338) prescription for good solutions, which are "typically local, rooted in a deep understanding of local histories and problems, so progress is likely to be the product of many small experiments rather than one grand plan".

In a similar vein, Moriggi et al. (2020, p. 285) argue "that by employing a care lens, practices become tangible and salient accounts of how transformations can be enacted in various realities. Their situational and contextual nature is thus an added value rather than a flaw". These accounts—vivid imaginings and experiences, told with emotion and from a perspective—are essential to cultivating the type of public emotion and projects that Nussbaum identifies as essential for aspiring liberal democracies. Such accounts must be situated and specific to elicit public emotion in support of liberal causes, such as sustainability. And they must be generalized to create a dialogue and bridge with more abstract principles (Nussbaum, 2015, p. 319). These accounts are also rooted in Care and a related feminist theory: *response-ability*, introduced by feminist philosopher Donna Haraway (2016).

Following Haraway's (2016) notion of "response-ability", this involves the capacity to not just answer for our actions but respond to something or somebody from the socio-ecological environment in which we are embedded. Such ability for responsiveness is not motivated by legal obligation, nor is rooted in relationships of blood. Rather, it comes from multiple practices of relationality: the more we engage in attentive relationships, the more we feel the need to care about and for others (Tronto, 2013, p. 288).

Some scholars worry that the ethics of care reifies traditional gender roles because it's founded on the concept of Caring at home. It also often fails to consider intersectional theory and the full diversity of homecare experiences. Here, I might note harmonies between Caring and ethical frameworks that had no discourse with feminist theory. In Confucianism, for example, we can identify a similar concern for interdependence, relationship, and societal obligations.

In his book *The Moral Sense*, James Q. Wilson (1997, p. 197) illumines the core values of the Enlightenment: "The generally shared corollaries of that belief were a commitment to sceptical reason, personal freedom, and self-expression. Elsewhere, communalism, tradition, and self-control remained the dominant ethos". While people in liberal democracies enjoy a freedom of expression unparalleled in the history of humanity, independence has become so stressed that individual citizens often lose sight of *interdependence*, a ruling ethos for so many societies from ancient times until today. Non-Western cultures did not undergo the Enlightenment philosophical movement, which elevated science and scepticism over received wisdom from ancient sources. Perhaps China provides the most preeminent exemplar of a non-Enlightenment culture; its culture traces back continuously over thousands of years, far into the Bronze Age. The Maoists shifted the cultural and educational model of China away from the gentle space of Confucianism and to a more stringent terrain, yet the culture is still continuous when contrasted with the West and its core values, which trace back not to the Bronze Age, but to the eighteenth century. Wilson (1997, p. 197) outlines a Chinese philosophical system that does not stress individual freedom: "For several millennia the ruling spirit of China was harmony and conformity, whether based on the flexible moral principles of Confucian thought or the rigorous and exacting legal codes of the Qin and Han dynasties, and the ruling doctrine was that of collective responsibility: families were responsible for the conduct of their members".

Three philosophies explain why Asian schools, and thus vastly different Asian societies, place such enormous value on interdependence. Confucianist, Taoist, and Buddhist philosophical systems have informed the Eastern consciousness just as the Enlightenment informed the Western consciousness. Confucianism also believes that the self is an illusion, but it offers a more pragmatic explanation for the illusory self. An individual defines herself only by the roles and relationships she plays, as explained eloquently by the eminent Confucian scholar Rosemont (2012, p. 9): "In order to be a friend, neighbour, or lover, for example, I must have a friend, neighbour, or lover. Other persons are not merely accidental or incidental to my goal of fully developing as a human being, they are essential to it indeed they confer unique personhood on me, for the extent I define myself as a teacher, students are necessary to my life, not incidental to it...Our task is to meet our responsibilities to others". As a foundation for learning, Confucianism corresponds with Caring and the response-ability. This distinct worldview shows that we need not stereotype Caring and response-ability with traditional notions of maternal care. There remain other perspectives in which to understand our interdependence.

Taoism, often translated as "the path", offers a complementary worldview to Confucianism and its focus on social responsibility and interdependence. Taoism stresses that the path to perfection remains within each person; one achieves harmony and perfection by doing what comes naturally, a way of living integral to childhood but elusive throughout the rest of life. Taoists "let it flow", like water eroding a stone, and seek balance between extreme states of emotion; indeed, this ever-shifting process becomes identity. Unmoored from a fixed identity and aware of interconnectedness, Taoism engenders a sympathy towards nature and all of life (see Lee, 2000). Taoism uses water as its metaphor for interconnection and worked with the subtle interplay between the particular and the general. Yet the general idea, the Tao, remains alive with spontaneity.

Case Study: Eddy at New Paltz

According to Photography Professor Andrea Frank, Eddy is: "an evolving, open, transdisciplinary, co-creative, and embodied learning lab. Our educational practices and systems support our current paradigm that incentivizes competition, oppression, and destruction" (A. Frank, personal communication, November 9, 2021) (Fig. 9.1).

Instead, she asks a How Might We question, often framed during the define phase of design thinking: "How might we, together, live, learn and educate around values and practices that are care-driven, relational, restorative, sustainable, and focused on the long-term wellbeing of all beings?" To answer this question, Frank and her co-creators reframed education using nature as a metaphor. On a practical level, the answer involves all the elements of Caring Education. At Eddy, participants model Caring, with an emphasis on listening well. They engage in non-hierarchical dialogue, with open-ended discovery as its goal. They practice, with design incarnating as an intent, a playful process, and a prototype. They confirm, affirming the best possible motives for everyone in its circle of trust and reciprocity.

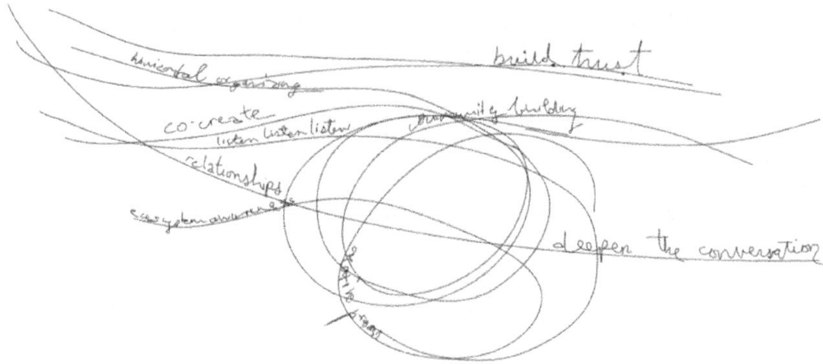

Fig. 9.1 Thinking through drawing an eddy

At Eddy, we also take part in the U Lab by the Presencing Institute. Founded by MIT Lecturer Otto Scharmer, this organization introduces leaders to the source of their intentions. In their book *Leading from the Emerging Future*, Scharmer and Kaufer (2013, p. 69) write: "Just like water in the physical system, the makeup of people in a social system stays the same under a given set of conditions. The difference between natural laws and the social field is that the actors in a social system are able to initiate change. In other words, they are sitting in the water while the temperature changes—and they can potentially get their hands on the temperature control. When their field state of awareness or consciousness changes, the actors relate to one another in different ways, and end up co-creating very different results".

Eddies at New Paltz include events such as *Sound Your Truth* (Fig. 9.2). This project evolved from playful collaboration and role-playing between campus faculty. Students from historically marginalized communities took centre stage on campus, speaking poetry, singing, and making artwork to vocalize their perspectives.

Ancient Greeks saw the design of such performative work as vital to democracy. In Athens, a tragedy could help people treat human vulnerability with compassion, and even objects of disgust could transform in the light of compassion. A comedy, meanwhile, could help inculcate a sense of lightness and playfulness in response to flawed humanity. Nussbaum (2015, p. 260) notes how these performative acts held deep emotion and at the same time were "important inputs for political discussion". If ethics

Fig. 9.2 Sound Your Truth: Thursday, October 14, 2021 (George Floyd's birthday) (bottom right David notMD, CC BY-SA 4.0 https://creativecommons.org/licenses/by-sa/4.0, via Wikimedia Commons)

is about eudaimonia, about how we might best live to promote mutual flourishing, then politics is about who can take part in the world and benefit from it. Through Caring Design, Eddy created the conditions for students to reengage with skills vital to democracy, including critical thinking and empathy, with its imaginative associations. The listening event, Sound Your Truth, has an ongoing Eddy in the Listening Sessions, which students and faculty continuously conduct with each other and share on the Eddy blog.

One of our major projects took place on April 27, 2022: *Health, Wellness and Environmental Justice: A Symposium on Healing*, with the nearby city of Newburgh, New York (Fig. 9.3). Led by Dr. Edward Lawson, SUNY New Paltz Assistant Director of Intercultural Relations and a lecturer in Black Studies, English Professor Sarah Wyman, and Dr. Erin Bell of the University at Albany, this symposium brought 50 to 75 students from Newburgh, grades six to twelve, to the SUNY New Paltz campus.

By health, wellness, and environmental justice metrics, Newburgh has suffered from the negative impacts of systemic racism. Recently, the city

Fig. 9.3 Newburgh Symposium: Wednesday, April 27, 2022

experienced drinking water contamination from PFAS, a dangerous chemical that lingers forever in a person's system, found in Teflon. At this event, our plan included students interviewing experts, creating posters pursuing big ideas and questions, community mapping, and the co-design of creative habitats that will be placed in Newburgh, as sites for creativity.

In accord with relational response-ability, our collaboration with Newburgh will be ongoing. Dr. Bell leads an ongoing public health effort in Newburgh, and teams from Eddy have embarked on listening sessions with community members in Newburgh. We intend to foster relationships of care that help to overcome Care's impediments. Passion and co-creation can help with the fatigue that stems from the inefficiency of Care: Eddy distributes response-ability to a group; though individuals might lead in their expertise, they also have people to support them or help them imagine new ways of working. The narrowness of Care can be remedied through symposiums and listening sessions between communities that might otherwise have seemed distant despite their geographic proximity. Listening and storytelling can help us feel each other's pain-points and aspirations, prototyping solutions that attend to deeper levels of human experience. Our work with experts in public health, meanwhile, will assure that we direct Care appropriately despite the powerful sway of vivid stories. Finally, a diversity of perspectives means Care can draw upon diverse cultures and traditions.

Conclusion

Typically, education focuses on disciplines rather than Caring relations. Yet even in a traditional twentieth-century design education that focused on disciplines, designers learned in an integrative way. Designers are already uniquely situated to be the type of caring, collaborative, interdisciplinary people needed to confront the complex challenges and heal ecological, social, and spiritual divides. Designers can be leaders in transforming communities from "egosystems" to ecosystems. In ancient Greece, ego meant "I" and eco, "house". We can think of this shift as being one of moving from competitive achievement, the pipeline, to the type of caring and cooperation one might experience in a loving, functional household.

As we transform design education so that it emphasizes a Caring purpose, design students can work on a foundation of philosophy and ethics. They can be aware of these discourses, to guide them in a world saturated with information and misinformation, and emotions that block compassion, which Martha Nussbaum has identified as fear, envy, shame, and even disgust. Based on my ongoing experience with Eddy at New Paltz, I recommend Caring and compatible worldviews from diverse cultures that elevate interdependence and relationships. When people are creative in a Caring space, they are more likely to flow ideas together—even if they run counter to the mainstream. Here, emotion can guide reason. Emotion can become an instrument for the type of dialogue that leads to discovery, and a bridge from individual stories of transformative change to a collective one. These are the Eddies, finely aware and richly responsible, that we hope to create.

Acknowledgements I would like to thank my Eddy core team colleagues—Andrea Frank, Michael Asbill, Sarah Wyman, Anthony Dandridge, Emily Puthoff, and Emilie Houssartat—at SUNY Paltz. Other people join this core team all the time, and there are many Eddies around Eddy Listening Sessions and campus events. Finally, I would also like to thank Eddy & Newburgh co-leads Ed Lawson and Erin Bell (University at Albany).

References

Hamington, M. (2019). Integrating care ethics and design thinking. *Journal of Business Ethics, 155*(1), 91–103.

Haraway, D. (2016). *Staying with the trouble: Making kin in the chthulucene.* Duke University Press.

James, H. (1934). *The art of the novel.* Scribners.

Lee, J. H. (2000). Finely aware and richly responsible: The Daoist imperative. *Journal of the American Academy of Religion, 68*(3), 511–536.

Moriggi, A., Soini, K., Franklin, A., & Roep, D. (2020). A care-based approach to transformative change: Ethically-informed practices, relational response-ability & emotional awareness. *Ethics, Policy & Environment, 23*(3), 281–298.

Noddings, N. (1991). *The challenge to care in schools: Rethinking the goals of education.* Teachers College Press.

Noddings, N. (2006). *Critical lessons: What our schools might teach, but do not.* Cambridge University Press.

Nussbaum, M. (1985). "Finely aware and richly responsible": Moral attention and the moral task of literature author(s). *Journal of Philosophy, 82*(10), 516–529.

Nussbaum, M. (2015). *Political emotions.* Harvard University Press.

Rosemont, H., Jr. (2012). Two loci of authority: Autonomous individuals and related persons. In P. D. Hershock & R. T. Ames (Eds.), *Confucian cultures of authority* (pp. 1–19). State University of New York Press.

Sanders, E. B.-N., & Stappers, P. J. (2018). *Convivial toolbox: Generative research for the front end of design*. BIS Publishers.

Scharmer, O., & Kaufer, K. (2013). *Leading from the emerging future: From ego-system to eco-system economies*. Berrett-Koehler Publishers.

Tronto, J. C. (2013). *Caring democracy: Markets, equality, and justice*. New York University Press.

Willis, A.-M. (2006). Ontological designing. *Design Philosophy Papers, 4*(2), 69–92.

Wilson, J. Q. (1997). *The moral sense*. Free Press.

CHAPTER 10

Design Challenging the Concept of Market

Tore Kristensen and Gorm Gabrielsen

Introduction

This chapter will explore contradictions between design and marketing by identifying links between design work and marketing work. The general image of the designer and marketing consultant is that the designer draws, uses computer aided design tools and makes material models, while the consultant works making computations by computer. Marketing has a long tradition of cooperating with creative disciplines, e.g. advertising (Bartels, 1976), however, most publications deal with marketing and other disciplines in a separate way. In this article, we attempt to get closer to the knowledge structures. A major issue concerns the concept of "knowledge", that is the nature of design knowledge versus the nature of marketing knowledge.

T. Kristensen (✉) · G. Gabrielsen
Copenhagen Business School, Frederiksberg, Denmark
e-mail: Tore@kristensen.mail.dk

G. Gabrielsen
e-mail: gg.mes@cbs.dk

M. J. Lehtonen et al. (eds.), *Design Education Across Disciplines*,
https://doi.org/10.1007/978-3-031-23152-0_10

The following anecdote can make the above appear more concrete: in the late 1960s, the US furniture producer Herman Miller in Grand Rapids, Michigan, invited the Danish furniture designer Poul Kærholm to its factory to explore the possibility of adapting his chair PK22 chair for the US, and possibly, wider international market (Robert Blaich, personal communication). The PK22 chair is made of steel and leather and has a distinct design. After four months, the project was aborted. The conclusion was that the quality standards of Kærholm were too high for Herman Miller; they could not produce and market the chair at a price the market would accept. That outcome may have been different today, but the whole exercise demonstrates how a very good design did not meet the assumed requirements of the market; the consumer was simply not willing to pay the price.

The next sections explain what we mean by design and marketing. Then we merge the concept to investigate what may occur. Finally, we explore how these issues may be dealt with in educational and pedagogical contexts.

What Is Design?

Design rests partly or traditionally on craft traditions (McCullough, 1996), where an artisan typically works for and develops an object, tool or artefact for a single user. This can be traced into modern digital design and typically happens in direct personal collaboration. Marketing originally started as personal selling and exchanges in physical markets and at that time had a personal relationship similar to that of design. However, since the early 1900s (Bartels, 1976) marketing gradually started transforming into a mass market activity, where general or aggregated economic models would become important (see also Alderson, 1965). Later, marketing would include psychological models and to a large degree concern itself with branding. The digital age, however, seems to have altered this practice once more.

To understand the concept of design, John Heskett (2005, p. 2) writes: "[d]esign is one of the basic characteristics of what it is to be human, and an essential determinant of the quality of human life. It literally affects everyone in every detail of every aspect of what they do throughout each day. As such, it matters profoundly".

Very few aspects of the material environment are incapable of improvement in some significant way through greater attention being paid to their design. Inadequate lighting systems, machines that pay scant attention to how people use them, badly formatted information are just a few examples of irritants that create cumulative problems and tensions. Even within this focus, the spectrum of possibility is still huge, and in this chapter, it will be necessary to focus on a limited range of examples, indicating only some of the range of approaches and choices available, rather than attempting a compressed coverage of the whole.

Our question, then, is how can we organize such a term "design" to provide a deeper insight into what design can provide to us from a behavioural point of view? For instance, what can we express as the values of design? This would deal with what good design can do for us as human beings.

Heskett (2005, p. 5) stresses the modification and alteration of environments in ways that "design, stripped to its essence, can be defined as the human capacity to shape and make our environment in ways without precedent in nature that serves our needs and gives meaning to our lives".

In this sense, environment is a very broad term. If we, for a first attempt, look at the environment as being an individual, that is the surroundings of a single person, in contrast to that of a group or crowd of people, we get a varied point of view. It makes sense to consider both individual environmental factors as well as those of a society. Another dimension may concern the materiality of these environments. The traditional way concerns looking at material environments, as John Heskett (2005, p. 6) suggests: "objects, communications, environments and systems that surround people at home and at work, at leisure and at prayer, on the streets, in public spaces and when travelling". In practice, design may at least consist of several types of approaches or schools, although they all seem to share the quality of practical problem identification and solving, using advanced drawing tools.

The original approach may be the craft approach, building on many years of practice, where the potential customer would ask a craftsperson (e.g. a smith or carpenter) to develop and build an artefact or object. In this case, the outcome is likely to be influenced by the physical dimensions of the customer. An axe or spade would fit the user who cooperates directly with the designer (e.g. Bibby, 1957). A much newer approach, usually associated with schools of art and architecture, is an artistic view. In this case, the user or customer is rather an image or a construct in the consciousness of the designer. Many such designers would have an education as an architect, for instance, and may be highly competent craftspeople. The third form is the engineering-based design. In this case, the approach builds on the engineering sciences.

Markets and Marketing

By market, we refer to an institution with access to a physical (or immaterial) place where demand meets supply of whatever goods we are considering. A typical market may offer several product qualities and prices, and those who demand may explore the qualities and prices and decide to buy or not due to a reduction in price they find acceptable. By marketing (e.g. Bartels, 1976), we refer to the producer or trader's approach to identify demand and "educate" potential buyers, where there is only a fragmented or no market. The matter can also be working on prices, distribution, selection of products and communication (e.g. Kotler & Keller, 2019). For our purpose, some overviews of what is considered as marketing knowledge see Bartels (1976) and Hunt (1991). Marketing has indeed a very long history as practical trading (Kotler & Keller, 2019), as such, it essentially rests on experience and practical knowledge. Marketing as "thought" is a much newer concept (Bartels, 1976). In the earlier days, much marketing rested on economic models with various extensions taking place later, including the focus on the concept of branding and marketing management.

Marketing knowledge is essentially a managerial approach to identify needs and wants and support the ability to develop and communicate company offerings to a selected market or segment of a market. A simple view may rest on cognitive models dealing with information more than knowledge (Johnson, 1987; Lakoff & Johnson, 1999). This means marketing, especially as a practice, depends on image-schemata and propositional models suggesting a simple view of an aggregated market, an image of the customer; how they feel and think about brands and a toolbox to use this knowledge to acquire income and profit.

Clearly, in an ideal situation there would be a closeness between design and marketing enabling both to integrate information and knowledge. If we consider in some details what the cognitive models can mean in this context, we will consider the four types, propositional models, image-schemata, metonym and metaphor (Lakoff & Johnson, 1999). Each of these has been given considerable analysis in the cognitive linguistics literature, so we keep it simple here.

Looking at design, a propositional model may refer to the characteristics of physical materials and perception rules. Traditional designs were largely made of wood, steel, leather, ceramics and other materials that would characterize the finished product. The production process would also typically be crafts based. With the advent of digital tools, this is changing and we may see further alternatives in the future. However, it may be that original crafts-based artefacts will be more expensive and be seen as a luxury item. That means of course that the cognitive aspect will change from essentially a hands-on cognitive model to a more academic one. Finally, metaphors are the new creative elements that can cover artistic expressions, technical solutions and other particular insights. The metaphor may be the real target, but it is grounded on the other cognitive models. That is how the design is embedded in tradition and embodied in the designer's experiences.

Looking at marketing, we find a similar, but a different set of cognitive elements. At the propositional level, we may find calculus models (markets and profit), like competitive patterns of existing situations, such as SWOT analysis. The marketing expert may use this creatively to identify opportunities and places in a market with less intensive competition. The image-schemata is possibly first of all the product life cycle, which sets the stage for many considerations. It gives an insight to the novelty, competitiveness and possible time span for a new design. A metonym may be branding and, more generally, competitive experiences. In modern marketing, the dominant metaphorical concept is brand concepts. Possibly, even without a brand issue, the product concept may resemble a brand.

When Design Meets Marketing

We propose that design essentially builds on embodied conceptual models; design builds on cognitive models such as image-schemata and propositional models. A hammer looks like an extension of the human arm and its problem-solving function may follow a physical rule. We will analyse aspects of design and marketing as information versus knowledge using the concept of justified true belief. To see how it may work, we shall consider the cognitive process of design in contrast to marketing, which considers the average or typical customer. Finally, we will set the stage for challenges to understanding the level of analysis, aggregation contra disaggregation. Together, these three issues will not solve the overall problems, but they will serve as an attempt to articulate a future research strategy through new concepts and approaches that build on existing cognitive models such as metonyms (existing forms) and new ones such as metaphors.

What Kind of Environments Are We Considering?

To deal with design and marketing, we need to consider the nature of environment in which such knowledge and practices are used for relevant problem solving. As indicated above, two aspects seem to dominate the understanding of environment; the level of analysis and development and the aspect of material nature. By level of analysis, we mean the focus on the individual versus the focus on the market or market segment (Lakoff & Johnson, 1999). With a history of being developed in a crafts tradition, design often focuses directly on the individual as the potential user and customer of the designer. In contrast, marketing builds on markets, which typically assumes an aggregate focus, while the reference to "the customer" often rests on an image of an "average customer". Moore (1991) is an example, where the reference is the customer interface system rather than the individual customer per se. Table 10.1 is an attempt to categorize environments that are all subject to human designs. It is simplified into dichotomies where continuous categories are the rules.

If we start from the left bottom, we find the most common aspects of design. These are individual or small-scale environments. Much product design is about creating bodily extensions, such as clothes, hand tools and furniture, for instance. A common denominator may just be product

Table 10.1 An overview of design and marketing environments

Aggregate (big scale) material environments	Aggregate non-material structure
Infrastructure	Social order
Landscape	Legal/moral rules
Climate change	National identity issues
Physical living conditions	Political processes
Individual (small scale) material environments	Non-material environments
Architecture	Branding (concepts and logos)
Retail design	Behavioural affordances and enabling
Bodily extensions	Personal identity (feeling "at home")
Artefactual environments (furniture, kitchen, tools)	

design. It is not always common to place architecture here, but the aim of architectural design is often the well-being of individuals, families and limited populations. Product design in general may follow the general rules of bodily extensions. This may also be the case for architecture, especially in the form of a monumental building. Going from the left to the right may today reflect, at least partly, the effects of digitalization. Even before digitalization there were different approaches to dealing with direct material issues versus more abstract issues of rules, power and symbolic and behavioural issues. Artefacts of archaeology or the craftsperson's identity would reflect this, as would local craft traditions.

When we aggregate this perspective, we may talk of an infrastructural environment. A nation or other groups of people live and adjust to climatic conditions in this landscape. We also expect the aggregate climatic values to set the stage for bodily extensions such as clothes. It is obvious that people living in a climate with cold winters differ from those living in hot environments and that this influences many important bodily challenges.

Design as Problem Solving and Meaning

When we apply Heskett's definition, several issues become apparent. Designs differ in how they provide balance between problem solving and meaning. At one extreme may be pure engineering design, where the design is concentrated on problem solving. Obviously, there must also be a meaning dimension, but it is often restricted to competent designers or engineers. On the other hand, there is pure meaning, which is best exemplified by an internal decoration. Many symbolic issues such as brand logos may in fact also qualify. See Lakoff and Johnson (1999) for further studies of symbols, images and cognitive models.

In ordinary life, people do not distinguish clearly between knowledge and information. The concept of knowledge is old, usually attributed to Plato. This ancient statement concerns knowledge as "justified true belief" (Gettier, 1963; Plato, 1992). Information may just be a mathematical or formalistic sign (Ashby, 1956) usable for many good purposes. However, in order to become knowledge it must also be true and be believed. Coming from a crafts tradition, design is closer to the knowledge issue due to the direct touch implied. Marketing on the other hand is closer to information, because constructs are to a larger degree constructed through calculus and estimation. There are discussions about

Table 10.2 Design as integral between problem solving and meaning creation

Meaning dimension	*Problem-solving dimension*		
	Low	*Medium*	*High*
Low Pure art, sculptures, paintings	Artistic design Fashion design Expressive design	Automobiles (e.g. Porsche, Ferrari Sports cars Sports equipment	Musical instruments (Stradivari, Gibson, Bösendorfer, Hammond)
Medium Interior, décor and other environmental conditioning	Classical "Danish Design" Classical industrial design	Hand watch Worktable	Personal computer (PC) (Apple) Vacuum cleaner (Miele, Dyson)
High Support of bodily function (hammer, hearing aid, heart transplant)	Clothes of all kinds	Extension of body (crutches); Tools (screwdriver, drill)	Toys for children

this, often referred to by Gettier (1963), who asks among other things, what if the "truth" is based on a lie, misunderstanding, etc. That issue is beyond this paper, but it should be underlined, that design seems to relate closer to knowledge than marketing in the sense that it is often based on the crafts of a person working for an individual client. This is a point we will return to later; marketing is more a matter of an aggregate issue dealing with different forms of numeric analysis, closer to treatment of information.

Table 10.2 provides some examples of how designs combine meaning-making and problem solving. This table is an attempt to connect issues in a continuous meaning dimension, connecting meaning as in art versus meaning in problem solving. An artistic approach may aim at alternating respondent's perception and feelings, while problem solving aims at changing the environments in some practical way.

The Challenges

In a sequence of the 1990s American TV drama "Frasier", Niles, the younger brother of the title character, sets out to study whether he wants to become a father by carrying a four-kilo bag of flour everywhere. Unsurprisingly, the experiment is a major failure, and Niles's father—a smart, retired police—says that the experiment failed because it was an intellectual study, not an emotional one, and then points to his heart. In a sense, this is a short version of the interaction taking place between design and marketing. Designers work with a lot of intuition when they try to understand, or maybe better comprehend, user needs and possible outcomes, while marketing to a large degree rests on concepts from economics and analytical psychology.

Essential in this reasoning is the extended mind (Clark, 2008; Johnson, 1987) which claims open connections between three things; the brain, the body and the environment. Clark (2008) writes of a cognitive leakage, which makes the connections between the three. This is "being there", opening for an isomorphism between cognition and material culture or in other words between brain, body and environmental artefacts, objects and other environments.

The Norwegian theologian and sociologist Eilert Sundt (1864/1976), who studied evolution under Charles Darwin in London in 1860, suggested a very interesting approach:

> Boat building is a joint project between builders and sailors (fishermen). A boat builder may be highly qualified, but still he cannot make two boats exactly alike, no matter how much he tries. The deviations that emerge must be called *accidental*. However, even a very fine deviation may be identified at sea; then, it is not accidental, and the sailors may identify the boat that has been improved or made more convenient. They would then advise the builder to follow up on these improvements in future boats to improve the experience and safety for others. (authors' own translation)

One Challenge Is About Levels of Analysis

As indicated, design historically emerged from craftwork. With the advent of the industrial era, interaction between makers and consumers dramatically changed. Now the industrialists and the designer had to meet the demands of the market more broadly. In addition to catering to an individual's needs, the challenge now included the production setup, and perhaps the distribution and market as well. The individual user therefore is often reduced to an "average" or "general consumer" or even a "stereotype". However, the challenge of identifying needs has become even more complex due to globalization and social media. The typical solution in the choice of media still assumes catering to an average or typical user or the largest reachable market segment.

In all social sciences, the ability to aggregate and disaggregate, that is switching between individual and societal levels is important. However, sociologists such as Durkheim (1895/2013, p. 83) claim it is nearly impossible a task and instead they argue that the proper level is the individual. The perception of sensory impressions like colour, shapes, emotional experiences (Damasio, 2000), facing new manufacturing, and other digital approaches enable and often require customized approaches. One only needs to check the many contributions to digital marketing (Kumar, 2018). To build on recent marketing development we suggest starting at an individual level and then aggregate, that is to apply a "bottom-up approach" in contrast to the conventional "top-down" approach. One might say this leads to the "stereotyping" of consumers. Big data enables a variation between consumers and sets of characteristics; however, it remains a market (and marketing) approach.

Discussion

In summary, we may repeat the point of "justified true belief", where justification may be formal as in (statistical) significance test or intuitive like that of an artisan. "Truth" may be a formal issue or an embodied issue. Finally, "belief" relies on experienced perception or expectations. Obviously, these issues are closely knitted together. Issues of marketing and design still may rely on different positions in this multi-dimensional mechanism or structure (Elster, 2007).

The cognitive models that marketing and markets rely on are to a large degree "image-schemata" and "propositional models". That means common and often visual references to brands, economic models and calculation techniques. Such models are often complementary to metonyms as visual expressions, to be perceived consciously or when they are parts of continual habits even unconsciously. The visual elements may just be naturally seized by the hand as a tool (pen, scissors or box of sweets) and as such closely related to the actual situation; need for writing, cutting or clearing the throat. One can imagine the references to commonalities, competing products, inspirations and elements, which are supposed to support user-interactivity, brand images and visual environments. Clearly sometimes, several competing cars or computers look so much alike that it takes an additional look to distinguish them apart.

Metaphors naturally represent the creative and new elements of design. However, for existing companies it is usually vital to follow some degree of likeness with the existing (metonymic) visual elements.

A moving body, such as the hand, may serve as an example here. The hand and the tool (e.g. computer) as a natural extension, and the connection we want to make, is the one between the moving hand and the cursor on a computer or iPad screen. Alternatively, we might have used a pen or something similar. By creating combinations of cognitive artefacts to fit or alter human cognitions and emotions is an important way of looking at the setup. In this case, the setup may simply be a complete workspace. A number of relevant tools for writing are present and the writer may not even think consciously about the choice.

Leading universities, in collaboration with schools of engineering, business and design, use cross-disciplinary teams to realize the need for cognitive models. Above all, realism in the teaching situation is probably the most important issue. Therefore, design should be at the core and artefacts, objects, drawings and combinations of, for example, image-schemata are urgently prioritized. It often takes place in elective courses open for both business, engineering and design students. Typically, the students are already well educated (e.g. they hold an undergraduate degree). The classes often take a particularly strong view of design but keep the other experts responsible for their roles: for instance, engineers find material and technical solutions, and business experts analyse demand in a selected market and calculate profit projections in a given situation.

A Few Pedagogical Implications

Classes in design and business in many universities including Aalto University, Massachusetts Institute of Technology, Stanford University and Illinois Institute of Technology (to mention a few) have all organized classes where designers, engineers and business students have worked in teams to conduct actual design work. In reality, the flow of information between disciplines in a tacit setting where imitation and learning is required is challenging. Exercises may be devised to give examples of differences between ways of deriving meaning from information across disciplines. Granted, information might change in the process as it is approached from different disciplinary standpoints, yet from such moments we can learn more about different disciplines.

The example of Poul Kærholm that we introduced in the beginning of this chapter is a prime example of how design and marketing disciplines are aligned in theory but not in practice. The real material issues of the design cost of production and expected demand may fit in a particular time and society, but not in others. It may also be that the more craft-based production of the furniture producer Fritz Hansen in Denmark, aiming at a luxury market; versus the industrial production principle of Hermann Miller, aiming at a larger market, did not fit very well. That could change over time or a different assignment altogether might have given a different outcome. Pedagogically this may call for some experiments, with simple resources. In any case, the learning might not have come from just reading about it.

Finally, a general issue concerns the lack of permanency in the environment. That follows from the examples too. A textbook in marketing or economics, for instance, often builds arguments based on constants. That is how many models and tools are developed. This does not mean they are worthless in decision-making. However, it is essential to consider the conditions on which they are dependent. The advantage of design is the real time, actual experimentation. Taken together, both design and business modes of working may gain considerable experience from collaboration.

References

Alderson, W. (1965). *Dynamic marketing behavior: A functionalist theory of marketing*. Richard D. Irwin.

Ashby, W. R. (1956). *An introduction to cybernetics*. Chapman and Hall.

Bartels, R. (1976). *The history of marketing thought*. Grid.

Bibby, G. (1957). *The testimony of the spade*. Collins.

Clark, A. (2008). *Supersizing the mind: Embodiment, action and cognitive extension*. Oxford University Press.

Damasio, A. (2000). *The feeling of what happens*. Vintage Books.

Durkheim, E. (1895/2013). *The rules of sociological method: and selected texts on sociology and its method* (S. Lukes, Ed.). Palgrave Macmillan.

Elster, J. (2007). *Explaining social behavior*. Cambridge University Press.

Gettier, E. (1963). Is justified true belief knowledge? *Analysis, 23*(6), 121–123.

Heskett, J. (2005). *Design: A very short introduction*. Oxford University Press.

Hunt, S. D. (1991). *Modern marketing theory: Critical issues in the philosophy of marketing science*. South-Western Publishing Company.

Johnson, M. (1987). *The body in the mind: The bodily basis of meaning, imagination, and reason*. University of Chicago Press.

Kotler, P., & Keller, K. L. (2019). *A framework for marketing management*. Pearson.

Kumar, V. (2018). Transformative marketing: The next 20 years. *Journal of Marketing, 82*, 1–12.

Lakoff, G., & Johnson, M. (1999). *Philosophy in the flesh: The embodied mind and Its challenge to Western thought*. Basic Books.

McCullough, M. (1996). *Abstracting craft*. MIT Press.

Moore, G. A. (1991). *Crossing the chasm: Marketing and selling high-tech products to mainstream customers* (1991, revised 1999 and 2014).

Plato. (1992). *Theaetetus* (B. Williams, Ed.). Hackett Publishing Company.

Sundt, E. (1864/1976). *På Havet*. Gyldendal Norsk Forlag.

CHAPTER 11

Blending Boundaries: Design and Technology

Derek Jones

Part II of the book is about blending boundaries in design and technology. For many design practitioners and educators reading this, working with boundaries may seem very familiar. Much of what we do in design necessarily requires us to work with boundaries that arise from roles, specialism, activities, cultures, and knowledge, to name but few examples. In addition, design is also distinguished by ***what designers do*** with boundaries: how we work across, between, at the edges of, pushing, pulling, or blurring them.

This notion of working with boundaries is arguably so ubiquitous in most design disciplines that it can often be taken for granted. By exploring what we do with boundaries as design educators, I hope to present a few interesting ideas about boundaries in design generally and relating to the chapters that follow in this section of the book. So, before you

D. Jones (✉)
The Open University, Milton Keynes, UK
e-mail: derek.jones@open.ac.uk

M. J. Lehtonen et al. (eds.), *Design Education Across Disciplines*,
https://doi.org/10.1007/978-3-031-23152-0_11

read about blending boundaries in the next few chapters, I would like to consider a really simple question: What do we actually mean by 'blurring boundaries'?

Crossing Boundaries

I would like to start with a recent example of a boundary that many educators encountered over the past couple of years and one that is the focus of Valencia et al. and Kornblat's chapters in this section of the book: the boundary between physically located and distance teaching. In response to the global pandemic, many institutions crossed the boundary from face-to-face to online and distance education modes, very often as a matter of urgency or emergency (Jones & Lotz, 2021; Winters, 2021). I use the word boundary here because, for most educators, the experience was a significant *modal* boundary transition, where one mode (physically collocated methods) was replaced by another (distance and online mediated methods).

In making the transition across this boundary educators met a range of unexpected outcomes and realised that simple moves (transference, translation, copying) between one mode and another is not always possible: what works in a physically proximate studio cannot necessarily be replicated or directly translated to an online or distance setting (Jones, 2020, 2022). Hence, there is something about the transition across this boundary that isn't obvious or straightforward: a boundary crossing worth considering.

Position/Starting Point

Before going further, I have to declare my own position. I teach design at The Open University (OU) in the UK. This is a national higher education institution set up in the 1970s to offer higher education at a distance and aimed at providing an alternative to the 'traditional' university model for students. Some kind of design has been taught at the OU since its inception, meaning we have been doing this now for nearly 50 years (Cross & Holden, 2020; Holden, 2009). Hence, we can say that it is, and has always been, perfectly possible to teach design at a distance.

One lesson from our experience is that it is not the case that physically proximate teaching is 'better' than online (or even of distance being better than traditional methods): they are simply *different modes* of learning and teaching. Hence, they require a different approach in terms of reframing how design education might work in those modes, not simply treating it as a boundary of difference (or even deficit). Hence, this is a boundary worth exploring in design education and, I would argue, blurring significantly (Lotz et al., 2019).

But before we get to reframing, it is worth starting simply and returning to the observation that some things that work in one mode fail to work in another. So, why is this the case? Why is it that some online practices work well, but others fail completely? What is it about our understanding of these practices that leaves us unable to work with them across boundaries in the ways we have experienced recently?

Face to Face 'vs' Online

As noted, the challenges of transitioning from physically proximate to online spaces have been the focus for many educators over the past years during the Covid-19 pandemic. In that time, educators have reported that replication of proximate activities does not work (Corazzo & Gharib, 2021), engagement seemed lower (Winters, 2021), and creating connections and communities was a challenge (Delen et al., 2021). All of these issues arise because they cross some boundary between contexts where something changes as a result. This was particularly obvious when individual elements designed for physically proximate settings were 'translated' to online and distance modes of working.

Trying to replicate studio at a distance and online is a challenge under any circumstance and many educators tried many different things, such as replacing studio sessions with asynchronous Zoom sessions or using online repositories to act as virtual pinup spaces, etc. What many educators found was that replicating synchronicity using online methods did not necessarily work as a replacement: time and synchronicity is not the only thing that has to cross the boundary for studio to work (Hepburn & Borthwick, 2021; Jones & Lotz, 2021). Or they found that replicating functions of traditional studio change these activities in unexpected ways (e.g. leading to different types of interaction or engagement) or they do not seem to replicate the experience of that same task in a traditional studio setting.

To generalise, in the boundary crossing from traditional to online modes, something happens to make elements work in unexpected ways. This happens for two main reasons.

Firstly, the things we are translating are not necessarily the ones that work across modes. The classic example is the attempted replication of university campuses in online settings, where the replication of 'surfaces' of buildings fails in terms of reproducing the original interaction, engagement, and use of these spaces. As any architect or designer will tell you, being in a place is more than just the surfaces and space (more on this later).

Secondly, we rarely translate 'enough' elements to recreate the scale of entity we have in our heads. We might think that to replicate studio it's only a case of replicating several smaller elements (such as synchronous meetups, spaces to share work, or messaging services). But in doing so it's easy to forget the true complexity of studio as a social, technical, educational, emotional, and transformational place of learning. The problem here is that studio is as much an embodied praxis and concept as it is a set of functions: a living idea of a continual and evolving praxis, meaning it's challenging, if not impossible, to 'count' the many parts it takes to make the whole.

So, if superficial replication doesn't always work, and if we're unsure of 'how much' is needed to translate a complex entity like studio, perhaps we can look at it from another perspective and ask what does work? What does cross such boundaries and remain the same, or work in a similar way?

You Are the Space

One simple response to the question of what is consistent across modal boundaries is: 'you'.

This answer appears trivially obvious, but it hides a much deeper idea: it is our ideas of the world (our conceptions) that we carry between modes and that are persistent across boundaries such as physical and online spaces (Jones, 2013, 2017). These conceptions are more than simple analogies or imaginings of the world; they are structures of thinking that embody our being and thinking in the world. For example, we know that recreating the 'surfaces' of space does not recreate the same lived experience: as noted above, a university campus recreated as a digital model does not necessarily allow the same experience of that physical campus (Addison & O'Hare, 2008; Gardner et al., 2008). Conversely, if the conception of space (place, phenomena, value, purpose, *genus loci*) is what we transfer, then it is perfectly possible to see the same rich, human behaviours in an online space that we see in physical ones (Clark & Maher, 2001; Minocha & Tungle, 2008; Twining & Footring, 2008).

If this is the case, then it must also be possible to work with conceptualisations of the world across boundaries such as physical/online spaces. This is quite a profound thing to be able to do and it involves a radical shift in how we approach the design of learning and teaching. Critically, it suggests that what we *think* makes a boundary, in terms of what is 'in' and 'out', might not be what we experience or conceptualise.

But what does this really mean? What practical or pragmatic use can this have for educators? One possible answer lies in how we think and, more importantly, how we do not think the way we think we think!

Mind vs Body

One way to respond to questions of thinking and conceptualisation can be approached using another boundary in this section of the book: that of mind and body. This is a particular focus of Kristensen and Gabrielsen's chapter in terms of exploring how this problematic separation must be put back together in design practice and education. Put simply, we still retain a Cartesian view of mind and body, especially in terms of what we 'think we think' by assuming that this is the same as what we are aware of (Damasio, 2006). This is an error of perception and, in reality, we are only aware of a very small fraction, around 10%, of our cognition (neural and cognitive activity). To put it another way, most of our cognition is inaccessible to us—we are unaware of it.

This remains an incredible facet of human cognition and consciousness and one rarely treated with the importance it deserves, and an interesting one to apply to design education. It could be argued that designers often operate in such interstitial or liminal cognitive spaces. Underlying concepts such as convergent/divergent thinking, problem/solution spaces, or any other mode of cognition that deliberately destabilises a thinking system (Ramage, 2017; Schön, 1971) is the ability to leverage cognition or cognitive states that we are less aware of or that rely far less on our executive function.

Our lack of awareness of our own cognition, I argue, helps explain the issues we face moving across boundaries like those outlined above: similarities we are unaware of suddenly become visible; differences we are aware of remain invisible. Critically, it supports the idea that there is persistence of conceptualisation—that we take our ideas of the world with us when we move between boundaries. This gives us a place to explore for answers, but conceptualisations of the world are far less tangible and harder to work with than certainties, such as the physicality of things or the functions of online spaces. Hence, we need to explore this area using the tools of cognition itself: the artefacts we use to think.

But what does this actually mean in practice? What practical or pragmatic use can this have for educators? Fortunately, the chapters in this section of the book give us examples of practical conceptual artefacts: metaphors, embodied conceptual models, and care, framed as human dispositions and capacities.

Metaphor, Embodiment, and Boundaries

As designers and design educators we know the power of metaphor to communicate and translate ideas. It could be argued that without metaphor we would be unable to communicate any ideas because it is effectively impossible to replicate the complexity of cognitive states between people. Hence, metaphors are well understood as tools to mediate communication in design, but what we might not realise is just how foundational metaphors are.

Lakoff and Johnson (1999) argue that such metaphors are constructed from our *embodied cognitive* experiences and that translations across concepts using metaphor are as physical as they are mental (albeit, such dualism becomes irrelevant when they are embodied). Examples of this are explored in Kristensen and Gabrielsen's chapter and their framing of 'embodied conceptual models' in design education. Another embodied conceptual metaphor can be found in the subject of this section in the book: boundary. A boundary is a definitive element. It bounds (surrounds, contains, circumscribes), meaning that it bounds something, creating two other things: an inside and outside. These terms are deliberate and preferential because a boundary must be something humans can conceptualise in terms of experience. Hence boundary gives us an inside (what can be bounded) and outside (stuff that is not inside).

Applying this definition, boundary is part of Lakoff and Johnson's famous container metaphor ('in') and it is critical to understand where this metaphor comes from to appreciate how it works in our minds. We do not learn what a boundary is because we are taught it or read about it; we learn the meaning of boundary from our embodied cognitive experience in the world. Our bodies are one of the first boundaries we experience and they are how we create the *concept* of boundary in the first place. We then expand this metaphor and apply it to other things that have similar conceptions, meaning we successfully apply this as a metaphor to things such as social, identity, or even abstract groupings (Lakoff and Johnson give the examples of being 'in' an organisation; being 'in' love). Even later, when we come across boundary as an idea expressed using other symbols (like this introduction), we still (re)apply our prior embodied conceptualisations of the metaphor. There is something quite foundational about the fact that we use so many spatial and relational conceptual metaphors, all of which are derived from simple physical experiences.

If we frame other subjects using boundaries, we will bring our embodied cognitive structure of that concept to bear on those subjects. We will create an 'in' and a 'not-in'; some delineation to represent the boundary; and we will treat things in or out, or that cross the boundary, accordingly. The boundary can sometimes dominate how we think about such moves, meaning we forget to acknowledge how these boundaries are created in the first place.

This same treatment of conceptual metaphor can be applied to the boundary created between traditional and online/distance design education. As noted earlier, if our conceptualisations can be persistent across such boundaries then we can make use of those and 'bypass' the boundary and work with it in another way. This is, I would argue, an interesting way of considering the term blurring boundaries.

Blurring Boundaries

If we retain conceptualisations of the world across boundaries then it's not necessary to *reconceptualise* when we cross boundaries. Instead, we need to *reframe* existing conceptions; to see the same conception viewed from a different perspective or viewpoint. That is, if the conceptualisation is persistent then this can become the focus of activity, not the boundary itself. Instead of treating the inside and outside as different, and the boundary as a delineator of that difference, a reframe focuses on the conception that persists, and then uses such differences as a different way to approach the same conception.

In many of the support sessions, the OU design school offered during the pandemic transition to online learning, helping educators *retain* existing conceptualisations of studio often led to the best outcomes. For example, instead of asking 'How can I get students to be together at the same time?', a question like 'What is it that supports moments of connection between students in my studio?' can be a more useful way to avoid the barrier of boundaries.

The difference in question here can also be seen from a pragmatic point of view. If we only focus on surfaces and obvious properties (walls, timings, actions, things), we ignore all the small but critical things that comprise the significantly complex set of properties that make up something like studio. Conversely, if we focus on the underlying values and all the small, valuable things that bring studio to life, it is perfectly possible to retain these as conceptions and ask how they might be supported using a different frame.

I argue that this definition of blurring boundaries is one of the mechanisms that makes it possible to move between modes of learning in design, as is shown in Valencia et al.'s chapter, demonstrating the direct use of conceptual metaphor to move across modal boundaries. Instead of facing boundaries head on and using them to frame difference, we can blur boundaries by usefully asking what are the underlying conceptions that stay the same.

Ethics of Embodiment

This brings us to another interesting consideration in embodied cognition and one that is often overlooked: values, ethics, and morality. If we begin with an embodied cognitive position, we have a completely different starting point for discourse on ethics or morality. Beginning with an embodied point of view, we must confront the 'bootstrapping' problem of trying to understand the very thing that is doing the understanding in the first place! To put it plainly, we are embodied thinkers, meaning our thinking, being, and values are hard to separate. How philosophers and critical thinkers have tackled this particular problem was a preoccupation of the last century and we have developed many new ways to approach these topics with an awareness of our own 'thinking positionality' (Jones, 2017; Lakoff & Johnson, 1999).

Again, setting aside the philosophy and theory, I would argue that such position taking and reflexivity in thinking (and doing) can be usefully considered as a designerly practice. It is another example of boundary treatment that explicitly finds a way to acknowledge the boundary without also simply falling back on treating it in the most obvious way. By moving across a boundary and acknowledging the change in position, designers 'blur' boundaries—at least in terms of boundaries constructed as irreducible and never-changing entities. Kornblat's chapter explores such alternatives to confronting boundaries, framing a form of ethics as care, expressed as a human capacity and relating this to it as a capacity in design. Once again, being a designer is to be an embodied being that holds ethical positions (regardless of what these might be) in a designerly way.

Back to Boundaries

As argued, moving from traditional to online education spaces is not simply a case of crossing a boundary—of looking over the boundary and asking 'what is different?'. The answers to such a simplified question can be insufficient, often do not work, and explain many of the failures observed in transitions over the past years. If all we do is move across boundaries, considering them in terms of differences and surfaces, then we will only ever replicate what is inside one boundary into another (i.e. extend an existing boundary) or, more likely, fail to replicate at all. However, when we approach boundaries in a designerly way, we can get different results.

Educators who took a design approach to the problem of transition experienced fewer problems simply because they brought a mindset of iteration, trial and (especially) error, and a *disposition* of working with boundaries in a designerly way. They blurred boundaries, making use of conceptions to extend existing practices into different modes. This is evidenced in the use of online tools for purposes they were not intended for, pushing them to their limits and beyond (Spruce et al., 2021); student-centred flexing of methods (O'Brien, 2021); self-critical reflexivity in practice and agile responses (Gray, 2021). Hence, I argue that there is a value to recognising how designers work with boundaries and in particular how designers blur boundaries. As designers, we rarely treat boundaries as hard limits and are trained to deliberately think 'out of the bo(x)undary'.

References

Addison, A., & O'Hare, L. (2008). How can massive multi-user virtual environments and virtual role play enhance traditional teaching practice? In *Learning in virtual environments international conference* (pp. 7–16). Open University.

Clark, S., & Maher, M. L. (2001). The role of place in designing a learner centred virtual learning environment. In *Computer aided architectural design futures 2001: Proceedings of the ninth international conference held at the Eindhoven University of Technology, Eindhoven, the Netherlands, on July 8–11, 2001* (p. 187). Springer.

Corazzo, J., & Gharib, L. (2021). Everyday routines and material practices in the design studio. *Design and Technology Education: An International Journal, 26*(4), 144–164.

Cross, N., & Holden, G. (2020). Design education in the open. *Open Arts Journal, 9*. https://doi.org/10.5456/issn.2050-3679/2020w10

Damasio, A. (2006). *Descartes' error: Emotion, reason, and the human brain*. Vintage.

Delen, I., Özüdoğru, F., & Yavaş, B. (2021). Designing during the pandemic: Understanding teachers' challenges in eTwinning projects. *Design and Technology Education: An International Journal, 26*(4), 182–204.

Gardner, M., Scott, J., & Horan, B. (2008). Reflections on the use of Project Wonderland as a mixed-reality environment for teaching and learning. In *Learning in virtual environments international conference* (pp. 130–141). Open University.

Gray, C. M. (2021). "Scaling up" and adapting to crisis: Shifting a residential UX studio program online. *Design and Technology Education: An International Journal, 26*(4), 37–49.

Hepburn, L.-A., & Borthwick, M. (2021). Synchronicity in the online design studio: A study of two cases. *Design and Technology Education: An International Journal, 26*(4), 71–85.

Holden, G. (2009). Design at a distance. In *Engineering and product design education conference*.

Jones, D. (2013). An alternative (to) reality. In M. Childs & A. Peachey (Eds.), *Understanding learning in virtual worlds* (pp. 1–20). Human–Computer Interaction Series. Springer. https://doi.org/10.1007/978-1-4471-5370-2

Jones, D. (2017). Embodied cognitive ecosophy: The relationship of mind, body, meaning and ecology. *Geografiska Annaler: Series B, Human Geography, 99*(2), 156–171. https://doi.org/10.1080/04353684.2017.1306971

Jones, D. (2020). Creating distance design courses. https://distancedesigneducation.com/creating-distance-design-courses/

Jones, D. (2022). *Studio use in distance design education*. The Open University.

Jones, D., & Lotz, N. (2021). Design education: Teaching in crisis. *Design and Technology Education: An International Journal, 26*(4), 4–9.

Lakoff, G., & Johnson, M. (1999). *Philosophy in the flesh: The embodied mind and its challenge to western thought*. Basic Books.

Lotz, N., Jones, D., & Holden, G. (2019). OpenDesignStudio: Virtual studio development over a decade. In *Proceedings of DRS Learn X Design 2019* (pp. 267–280). https://doi.org/10.21606/learnxdesign.2019.01079

Minocha, S., & Tungle, R. (2008). Socialisation and collaborative learning of distance learners in 3-D virtual worlds. In *Learning in virtual environments international conference* (pp. 216–227). Open University.

O'Brien, D. J. (2021). On country—Off country. *Design and Technology Education: An International Journal, 26*(4), 10–19.

Ramage, M. (2017). Learning and change in the work of Donald Schön: Reflection on theory and theory on reflection. In D. B. Szabla, W. A. Pasmore, M. A. Barnes, & A. N. Gipson (Eds.), *The Palgrave handbook of organizational change thinkers* (pp. 1–15). Springer International Publishing. https://doi.org/10.1007/978-3-319-49820-1_57-1

Schön, D. A. (1971). *Beyond the stable state: Public and private learning in a changing society*. Temple Smith.

Spruce, J., Thomas, P., & Moriarty, S. (2021). From sharing screens to sharing spaces. *Design and Technology Education: An International Journal, 26*(4), 96–111.

Twining, P., & Footring, S. (2008). The schome park programme—Exploring educational alternatives. In *Learning in virtual environments international conference* (pp. 366–377). The Open University.

Winters, T. (2021). Emergency remote studio teaching: Notes from the field. *Journal of Teaching and Learning with Technology, 10*(1). https://scholarworks.iu.edu/journals/index.php/jotlt/article/view/31580. Accessed 6 Nov 2021.

PART III

Shifting the Mindset: Design as a Catalyst for Rethinking Interdependencies Across Boundaries

CHAPTER 12

Teaching Change by Learning Change

Dilys Williams

Introduction: A Questioning that Informs Questions

> *In the twenty-first century, the well-being and even survival of humanity will depend crucially on our ecological literacy.... not only the intellectual understanding of the basic principles of ecology, but also the deep ecological awareness of the fundamental interdependence of all phenomena...as individuals and societies, we are embedded in and dependent upon the cyclical processes of nature.* (Capra & Luisi, 2014, p. 291)

Education is bound up in the interdependence of all phenomena and involves interactions between the environment and the laws of nature, cognition, and the laws of social systems and self-organisation and the cyclical logic of the self (Maturana, 2006; Varela et al., 1974). Tutors go through cycles of action and reflection to nourish and develop their own

D. Williams (✉)
Centre for Sustainable Fashion, University of the Arts, London, London, United Kingdom
e-mail: d.williams@fashion.arts.ac.uk

M. J. Lehtonen et al. (eds.), *Design Education Across Disciplines*,
https://doi.org/10.1007/978-3-031-23152-0_12

patterning of concepts, values, perceptions, and practices and to develop them with others to create a lens on reality. This forms the sense-making within a college, course, or other community of learners, that becomes an explicit or implicit means to recognise contributions to knowledge and understanding. Fashion education, with its vocational, disciplinary, socialising, critical, and creative lens on the world, has cycles that intertwine with those of nature, industry, culture, and society. However, it is evident (IPCC, 2022) that human activities are not in flow with the wider cycles in which they take part. Through extractivist and exploitative practices, our social paradigm is out of kilter with our ecological, and fundamental level of life.

This chapter focuses on a subset of these cycles, that of fashion design education, and how, by addressing misalignment at the level of tutor reflections and actions, we might leverage change that ripples across wider scales. It is informed by research by and with fashion educators over a three-year period, through a longitudinal co-inquiry involving researchers and practitioners in four universities in Europe and through research involving tutors in more than sixty universities around the world.[1]

If we are to serve students' and wider societal needs, we need to use the distinctions of our skills as design researchers and tutors to find ways to simultaneously unravel and knit practices differently to create new understandings and new forms of fashion and other creative practices. As a fundamental indicator of our identities, fashion's forms and understandings can permeate how we represent ourselves in the world, the conditions for livelihoods for millions of people, and manifest an honouring of nature's cycles. Education has a vital role to play in shifting fashion's cycles from a partial view that is focused on objects and materials in a consumerist, reductionist, economic growth paradigm, to a more expansive view of a gloriously entangled, interdependent set of activities and situated learning with material, human, and other resources at hand.

To navigate and develop world-relevant teaching practices, tutors must consider why they are teaching at all; what specifically they are teaching; how they are teaching, and learning, and who is involved? Looking into the core of the why, what, how, and with whom of teaching involves an intertwining of practice, tacit knowledge, and theoretical concepts.

[1] The outputs of the project include accessible tutor-resources, which can be found on an open source, online platform on the FashionSEEDS website, https://www.fashionseeds.org/.

This trilogy, and the systems-related writings of environmental anthropologist Gregory Bateson (1972) and philosopher Felix Guattari (2014) offer ways to connect ecological, societal, and personal perspectives on teaching beyond the explicit symptoms of mankind's ecologically and socially destructive practices. In the case of fashion, landfill sites piled high with discarded pieces and alarming emissions statistics come immediately to mind. Bateson (1972, p. 495) understands that "this massive aggregation of threats to man and his ecological systems arises out of errors in our habits of thought". He posits that a reductionist, mechanistic, rational worldview is the basis of an epistemological error. By recognising the errors in habits of mind, we can approach errors in habits of design, production, consumption, and indeed habits of dress.

Exploring habits of thought to change fashion's systems of teaching and of professional practice involves consideration of belief systems, world views, values, and intentions. The climate crisis is recognised as a crisis of perception, an existential crisis, and recognising the epistemological error of extractivist economies is difficult; "our everyday lives are rung with linguistic references that enforce its supposed legitimacy, through objectivist descriptions" (Lakoff & Johnson, 1980, p. 200). We should not underestimate the task at hand in developing new ways of knowing, learning, teaching, and living well together. The intertwining of mind, society, and environment is more than a surface-linking exercise; to get to the root of things, to foreground our ecological system, means exposing the blinkers of postmodern capitalism. The shared nature of the environment that we live in, and our collective impacts on it in emissions and biodiversity loss terms, reveal the commons on which all life and lives depend. In an industry-related discipline such as fashion, there is an opportunity to explore the tensions at play between societal, ecological, and personal living. Fashion and other design practices link the head, hand, and heart of matters, as well as creating connections between human, global and ecological scales. To transform education, so that it can be transformative, consideration must be given to and across these scales. "The only true response to the ecological crisis is on a global scale, provided that it brings about an authentic political, social and cultural revolution, reshaping the objectives of the production of both material and immaterial assets" (Guattari, 2014, p. 28). From this assertion, we can understand the need to change the intentions of education, as well as its content, pedagogy, and assessment criteria.

For the purposes of this study, we draw on Bateson's (1972) and Guattari's (2014) understanding of ecology as far more than concern about climate change, to see ecology as an epistemological system based on nonlinear systems governed by feedback loops and nonlinear causality. This calls into question what we value, why and how, to recognise prosperity in social, cultural, environmental, and economic terms. This, in turn, questions where knowledge comes from, who holds power, and how and with whom we teach and learn. This re-evaluation of thought requires that we give attention to an understanding of ourselves, organisations, disciplines, and industries we are part of, societies we live in, and the ecosystem we inhabit, as different scales of ecology.

The FashionSEEDS project uses a framework of fashion and sustainability developed by the author (Williams, 2019) based on a recognition that ecological, cultural, and social orders are the context of the economic order of things—and not the other way around. This framework foregrounds the role of culture in the creation of habits of thought and practices of design. "Now more than ever, nature cannot be separated from culture; to comprehend the interactions between ecosystems, the mecanosphere and the social and individual universes of reference, we must learn to think 'transversally'" (Guattari, 2014, p. 71). As educators, we need reflective spaces in which to ask questions of ourselves and each other, along with methodologies, pedagogies, and guides to apply what we know to how and what we teach. To understand *how* we know what we know involves moving beyond an overly rational, solutions-based approach to fashion education, into a multi-dimensional one, where ideas are emergent, context is ambiguous, and practices are relational. As Alfred Korzybski (1933, p. 247) said, "the map is not the territory".

The ideas presented in this chapter are drawn from my practice in leading a community of designers, researchers, and tutors at Centre for Sustainable Fashion, a University of the Arts Research Centre, based at the London College of Fashion. As a designer and researcher, I connect ideas, theories, and materials from many places, which I juxtapose to create collages of imaginings and doings. Fashion Education for Sustainability (FEfS) sits in the intersection between Education for Sustainability (EfS) and Design for Sustainability (DfS). Resonance between what we have been doing at a disciplinary level and education wide level can be found in the four areas of competences for educators: learning to know; learning to do; learning to be, and learning to live together, originally outlined as the four pillars of education (Delors, 1996). In presenting ideas about *how* we learn to know, do, be and live together, the chapter makes a case for an expanded approach to fashion pedagogy which recognises the need to facilitate a space for tutors' own reflection and action (as per Shulman, 2004).

Participants, Ambitions, and Hopes

Whilst there is a burgeoning network of fashion and other design educators and researchers seeking and enacting change, this study is the first, to our knowledge, longitudinal co-inquiry into fashion education for sustainability teaching practice involving tutors, researchers, programme leaders, and employers from more than 60 universities and 60 businesses around the world. This co-inquiry was made possible through the support of the Erasmus + Strategic partnerships and responds to societal and localised needs in the four countries of the partnership—Estonia, Italy, Denmark, and the UK. Each partner university brings distinctions not only in geographic terms but also in terms of university ethos, history, structure, and size. By cross-referencing different understandings and perspectives, the study sought to develop resources for thriving through education for sustainability with relevance in and beyond the partner locations.

The project's ambition was to create a heuristic for fashion education, where tutors act as "agents in the process of discovery" (Jickling & Sterling, 2017, p. 11). Through their experiences, ideas, visions, and doings, drawn from a range of source-types, tutors are well placed to act as the intermediaries between imagination and praxis. An education that recognises our ecological and social context is an education that is anticipatory, emergent, and participatory (Sterling, 2021). It takes humility to recognise how much we do not know about how to contribute to the world; it is humanity's arrogance that has led to our current predicament (Ehrenfeld, 2019). Whilst we have developed breath-taking technologies that enable us to do things unimagined in the twentieth century, we have yet to learn the fundamental lessons of how to live well together on a finite planet (Escrigas, 2016). In the words of Martin Luther King (1964, paras. 3) "We have learned to swim the seas like fish, and fly the skies like birds, but we have not learned to walk the earth like brothers" and, I would add, fellow earth inhabitants. This is a provocation to art and design educators that is both profound and exciting, as it invites each one of us to get involved.

Fashion, design, and sustainability education and practice involve sources of knowledge that are felt, passed on, and experienced; they do not always start with recorded and referenceable theory. Polanyi (1962) describes a pre-logical phase of knowing that can be visual, conceptual, and channelled through mindsets that involve communicating and sharing understanding. It is in this way that craft tradition and other skills have realised discoveries that are not published in peer reviewed journals and books held in libraries. Designers, tutors and students in art and design work both forwards and backwards, from a doing to a reading, from a practice to a theory. This practice-based approach to learning often invokes ideas better communicated through sketches, notes, photographs, or reciprocal interactions. It is thus important to recognise attendant learning that is social, responsive, and drawn out of the actions and ideas of tutors and students, alongside the more usually recognised structured learning that is intentional, formal, drawn out of policy, and validated through explicit structures and systems.

Questions Emerging from a Questioning

The long-term aim of this research is to develop practices that contribute to an equity and earth-centred ethos, infrastructure, and culture in fashion and wider art and design education. The immediate aim is to highlight and support the critical role of tutors as participants and intermediaries in change-making. University is simultaneously a space for students to acquire practical skills to apply in their livelihoods, and a space for students to discover novel ways of thinking and transform themselves in a more innate manner. A tutor's presence in the tension between these strands creates great possibility. To make the research usable and relevant to as many tutors as possible, research objectives included compiling data from surveys, interviews, and desk-based research to create a benchmarking report, the development of resources for tutors and an open-source platform to share project findings. These objectives were framed within the objective to develop and apply a framework for teaching fashion and sustainability with a series of underpinning pillars, a range of pedagogical approaches, guides to increase tutor agency, and a range of reference points for recognising change. The study was guided by the following two questions:

What conditions and resources do fashion tutors need to be able to change curriculum, change themselves and demonstrate the value of fashion and wider art and design practices to climate and social justice?

What kind of learning and change is needed, so that equity and earth-centred practices are more likely to be fostered and amplified amongst tutors and students?

Research Design

Having identified tutors as the focus of the enquiry, a mixed methods approach to research design was developed, using a co-inquiry methodology across the research team. This involved an extended dialogue with project team members across the project lifetime, running alongside the rest of the research phases. The longitudinal element of this research involved workshops and shared reflections with four to six team members from each of the four partner universities. This emergent phase of learning with and from each other's experiences as co-subjects and co-researchers enriched and informed the enquiry, using a methodology drawn from the extensive co-inquiry writings of (Heron & Reason, 2006). This ongoing reflection and action created content that was prototyped, tested, and refined across the accompanying research phases with a network of tutors and students from the four universities and across the project's network.

The research started with a mapping of fashion education and sustainability practices and needs, with reference to the project's core concepts of the four pillars, seven pedagogic principles, and three scales of transformation, using online questionnaires, interviews, and focus groups. The focus of the project benchmarking was to gather data from tutors in universities across the European community, in keeping with the ambition to contribute to regional sustainability. The project team was able to extend the reach of the survey to include responses from tutors in locations around the world, to at least partially fulfil the ambition to learn from practice beyond western based institutions. Thematic and granular findings were synthesised through the application of qualitative and quantitative data analysis to produce a Benchmarking Report (Williams et al., 2019).

Findings from the benchmarking phase were applied to the development of a Framework Prototype, tested through an immersive, co-learning experience with a group of tutors from across the four partner universities. New findings were drawn from these workshops via feedback relating to content, process, and tutor needs and interests. Prototype content from a parallel Materials Development workstream informed a review of existing and project-specific resources and tools of relevance to participatory learning, design, fashion, and sustainability. A further research phase included questionnaires and focus groups involving potential employers, Non-Governmental Organizations, government departments, and civil society organisations, to develop a Future Skills Foresight Document to inform the project's development (Williams et al., 2021).

Developmental phases involved the collation, design, and testing of resources of relevance to tutors across locations and teaching levels, based on the project's core concepts and research findings. These resources spanned the teaching cycle from unit or course ideation to quick reference points for use in a workshop or tutorial group. The final phase of research and development included a review of the trialled pedagogies and of the longitudinal co-inquiry, informing the development of a framework for reflection and action.

Insights from Benchmarking Review

The findings from the research phases were rich and plentiful. Whilst mindful of the value of the diversity of insights, perspectives, practices, and identified needs, this chapter seeks to focus on findings relating to pedagogy and practices that can amplify radical action towards teaching and learning as sustainability in personal, professional, societal, and biosphere terms.

Connecting and co-learning: there is a strong interest in embedding sustainability into student experience, curriculum design, interdisciplinary partnerships, and institutional values. This extension of the role of the tutor beyond their own courses and student groups into becoming actors in whole systems change includes an ambition to dissolve the academia's siloed ways of working and to increase synergistic relationships within and between academia, industry, and society.

A deepening of knowledge: there is an appetite to draw on a range of knowledges and to take a multi-faceted approach to pedagogy. Engagement in each of the seven proposed pedagogies was identified, with creative and critical thinking ranking highest, whilst informed decision-making through reference to science-based data and place-based learning were least widely explored, evidencing gaps in situated learning and interdisciplinary research.

Bridging thinking into doing: Culture is at the heart of creative practice and of the four pillars of sustainability, culture is recognised as the area in which fashion tutors are making the most valuable progress. Environmental exploration comes second, whilst consideration of economic parameters of sustainability is the least well-explored. Tutors openly seek to expand ways to influence change through extending opportunities for co-learning, especially in developing their own knowledge and practice.

A shared and empathic understanding: there is a recognition of affect in relation to risks and opportunities that tutors, students, and businesses have in taking action. Reciprocal processes of teaching and learning must take into consideration the contexts in which learners find themselves—or are preparing themselves to encounter. Tutors can create safe spaces in which to explore tensions between what is and what might be, in terms of sustainability.

Expanding horizons of time: learning is an ongoing process and there is a need to reach beyond the parameters of current course timelines and formats. Life-long learning, cross-generational learning, in-work learning, and skills transfer can all be informed by and inform undergraduate teaching and learning. Expanding the flexibility, adaptability, and nimbleness to change modes of learning is vital in creating a more inclusive Art and Design Higher Education system.

Questioning consumption and growth: at the core of aspirations of tutors lies the ambition to explore concerns about accepted logics and normalised behaviours that permeate educational goals as well as industry and government agendas. The role of design research in transforming how we live, work, and recognise our contribution in the world must be recognised, and a need to better articulate what creative practice offers is seen as vital to our futures—and to creative education's ability to thrive.

These thematic findings informed the development of the research outputs which integrate the pillars, pedagogies, and levels of transformation into a set of resources and guides. The draft platform content was tested through a further round of data gathering from interviews and focus groups, leading to the addition of an eighth pedagogic principle. Each of the principles are outlined as follows:

Insights from Trialling Pedagogies

Taking a constructivist approach that understands fashion design for sustainability as possibility-creating as well as problem-solving (Williams, 2015), the pedagogies trialled through this project were developed out of longstanding teaching of fashion and sustainability at undergraduate, postgraduate, and in-work training levels of study. The project applied these pedagogies to workshop content and courses, curriculum and learning design, adapted, and expanded in response to the research findings. The pedagogies are based upon the three dimensions of signature pedagogy (Shulman, 2005, pp. 54–55): "surface structure, deep structure, and an implicit structure". These dimensions cover the operational "what" of teaching and learning, with understandings relating to "how" to put pedagogies into practice. For tutors to reflect on and discuss moral beliefs relating to values and professional attitudes involves the "why" that is often squeezed out of staff and course development time.

Systemic Thinking

This is a foundational pedagogy in fashion design for sustainability that encourages a recognition of the living systems of which we are a part. It supports integrative and adaptive processes of thinking and practice that are integral to education for sustainability (Ryan & Tilbury, 2013), which, at a deep structure level, involves exploring relationships between people, places, and environments. At an implicit level, it draws on participants' belief systems and perceptions of interbeing (Hanh, 2017) in the world, whilst at surface level, it is about drawing feedback loops to visualise the cause and effect of actions.

Creative, Critical Thinking

This is a foundational pedagogy in art and design, where learners question and explore new and distinctive ways to extend the scope of a discipline or practice itself. At a deep structural level, this involves questioning structures in society and forms that fashion can take, at an implicit level it is a searching for self in relation to other human and more than human lives and related power dynamics, whilst at surface level it takes the form of diaries and sketchbooks reflecting thinking and action.

Participation

Participation is part of an underlying ethos of inclusion, belonging, and equity that actively seeks to eliminate discrimination and barriers to learning. At a deep structural level, it is part of a "doings of care" (Puig de la Bellacasa, 2010, pp. 162, 164–165) recognised by "activity that maintains and regenerates our world". It involves emancipatory education, education as praxis, drawing on wide sources of knowledge. At an implicit level, it is the elements that create agency for voicing concern and taking part and at surface level, it is in the wording of briefs, reading lists, and references used in class activities.

Informed Decision-Making

This pedagogy is an approach to identifying, analysing, and applying valid data into the creation of products, services, and systems. It presents a challenge to art and design education, which does not specialise in complex quantitative data analysis. At a deep structural level, informed decision-making is connecting trusted, meta-level data to situated contexts (connecting across system levels). At an implicit level, it is the discourse between disciplines as a co-inquiry that explores different disciplinary practices. At the surface level, it is a decision-making process to analyse the validity and applicability of data available.

Futures Thinking

Futures thinking integrates signature experiential learning pedagogy with speculative design practices, drawn out of interdisciplinary literature and design practice, notably referencing the work of Dunne and Raby (2013). At a deep structural level, it challenges the ontology of fashion activities through multi-sensory, multi-modal, multi-timescale approaches to design. At an implicit level, it expands the horizons of time, it invites thinking beyond the human and current socio-economic structures of production and consumption. At the surface level, it involves methodologies such as scenario planning (Wilkinson & Kupers, 2013, pp. 118–127) as well as a range of literature in inclusive design and transformation design.

Place-Based Learning

Place-based learning connects the signature pedagogy of situated learning and localism. It can be practised in teaching and learning from early childhood to lifelong learning environments. At a deep structural level, it contextualises design within a community in environmental, cultural, and societal terms and involves learning that is embedded within activity, context, and culture. The dynamic, non-static, emergent dimensions of this learning involve unintentional rather than deliberate aspects (Lave & Wenger, 1991). At an implicit level, it involves moving beyond the classroom and engaging in community, which at surface level includes fieldwork and partnership projects that extend beyond the confines of fashion networks and uses "practices of deep listening, direct observation, and multi-sensory data capture" (Williams, 2018, pp. 567–568).

Interdisciplinarity

As a pedagogy, interdisciplinarity is the consideration of how learners with different approaches, methods, and processes, in different disciplinary, cultural, and other settings can learn with each other. At a deep structural level, it's about hybridity, where education avoids essentialism, pure identities of nation, race, or role, extending capacity for understanding multiplicity and diversity in knowledge production. At an implicit level, it is pluralism in approach and perspective to offer a rich learning opportunity, a discourse with multiple perspectives. At the surface level, project-based learning, a longstanding art, and design pedagogical approach lends itself to the intertwining of ideas from a range of disciplines, cultures, and perspectives.

Learning Through Making

Making as a location and practice of learning is a foundational pedagogy in fashion, and wider art and design education. At a deep structural level, this involves equity for contributors as the basis for mutual learning, explored through traditions of making, including craft and indigenous knowledge. At an implicit level, this is about learning to know, learning to do, and learning to live together (UNESCO, 2011, p. 6) and at the surface level, it is about developing capabilities and skills for livelihoods. Whilst a fundamental part of art and design practice, it was not identified as a pedagogy in the first round of research but added in response to the co-inquiry findings.

The three major barriers to being able to realise these ambitions were identified as a lack of time, lack of relevant resources, and a lack of staff development opportunities. These conflate, with workload pressure offering little time for research, high-volume, low-quality resources, and little recognition of the importance of dedicated staff development time and spaces for interaction with peers and others outside of their institutions. The gap in tutor learning and reflection opportunities, environments, and resources for deepening understanding and its implications were identified as concerns in a rapidly changing industry. Tutors expressed that learning opportunities would increase their confidence in teaching and thus their agency in change-making and supporting learners as change-makers.

Changing Ourselves as We Guide Others: Findings and Learnings

Through the semi-structured conversations about our own practices, the project allowed for the sharing of experiences and ideas through convivial interaction. Evidence demonstrated that these sessions supported a sense of shared purpose and increased confidence in work done outside of the project. The process was not always comfortable or easy, with team members at different stages in their careers, and different institutional expectations. The observations of the longitudinal study chimed with those of the research findings of the first and second round of data collection and it became evident that tutors respond positively to opportunities to share experiences with each other, and that purposeful practice change is enabled and enhanced when tutors work together. These findings led to the designing of resources that are based on semi-structured peer-to-peer interaction with the aim of enhancing and amplifying tutor reflection and action cycles to deepen knowledge and agency. Alongside the development of course design, content, pedagogies, and assessment, a final element was developed, "The Fashion Tutor as Sustainability in Practice", is a three-part resource that can be engaged with separately or in tandem with each other specifically to recognise and support tutors own learning.

The process of developing these resources has been iterative, based on real-life experience with reference to theory and practice in co-learning and reflection including Heron and Reason (2006) and Murray (2011).

This three-dimensional approach to change also draws on transformation design, and Education for Sustainability (Sterling, 2009), Design for Sustainability (Walker & Giard, 2013), and meditation practice and ecological writings (Naess, 2008). These guides seek to gather tutors together to consider change as a dual (but not dualistic) process taking place within themselves and within learners and world in which they interact. By making these practices explicit, it is hoped that they and other practices of co-learning can be recognised as part of what is valued in teaching and thus included in work plans.

Ways of Being

A framework was developed to create conditions for a shared learning experience. It offers a light touch guide usable in informal and movable settings. It takes a diary format, for two people to engage in listening, reflecting, and taking action in ways that are supportive and enjoyable. What is shared and the intention of the process is decided upon by the participants. This mutual learning process where two people come together to develop their own sense-making via discussion, decision-making, and care is not limited to fashion tutors and does not limit the conversation to teaching, but rather explores the situated context of the tutor, recognising that self-care is intertwined with the care of others and wider activities that replenish life and lives, human and beyond human, in an interdependent world.

Ways of Knowing

A second framework was created for group reflection and action, drawn out of the literature review on co-operative inquiry, prototyped for use in the longitudinal study of the project and adapted again to be applicable as an open-source resource. It references "By the Fire", a tutor co-learning project led by Liz Parker and Lizzie Harrison (2016), and other learning design programmes. It involves inquiry comprising an intentional interplay between reflection "sense-making" and action and integrates four ways of knowing: experiential, presentational, propositional, and practical.

The resource responds specifically to the findings from the study relating to the need to "make time" to learn from others and to find resources that can be trusted and are relevant. It is also designed to be integrated into workplans, making it accessible to tutors who are full time, part-time or visiting tutors working in a range of roles. By making the process explicit, it seeks to give recognition to the process of tutor as a learner, encouraging teams and management to plan co-learning into work plans. The practice of co-operative inquiry is a way of working with people who have similar concerns and interests to oneself to understand the world and develop new and creative ways of looking at things and to learn how to take action to change things and find ways to do things better. (Heron & Reason, 2006). This methodology is particularly apt for fashion and sustainability learning, as it encourages a range of sources of knowledge, reference points, and inclusion of ways to share tacit knowledge, non-text contributions, including experiential elements of knowledge sharing.

Ways of Doing

The third element of the resources for tutors involves a less explicit approach to condition creating. Through developing a user journey or navigation system for the online platform, it encourages tutors to take routes most suited to their needs. The ways of doing route-mapping and content creation respond to the research findings that, at times, tutors need the immediacy of materials that can be nimbly applied. It is important for tutors to be able to have resources that match the pace and stage at which they are working. The user design means that the resources can be approached in multiple ways, thereby acting as a guide that can be referenced, adapted, applied, and used to gather teams together in transforming what it means to teach and learn fashion in an ecological and equity context.

In Conclusion: Foregrounding Cultures Changes Everything

Environment, culture, society, and economy form the core parameters of this project, threaded into workshop content, course, curriculum, and learning design and in the consideration of pedagogies and peer and group reflection guides. Whilst many sustainability frameworks recognise three of these four pillars, a mindedness towards culture is as critical as environmental, social, and economic considerations. The word culture is highly contested; its interpretation in the English language alone has many complexities as it involves deeply held, non-verbal, as well as verbal signs. Culture is ways of life; customs, beliefs, and rituals; codes of manners manifest in dress, photography, ceramics, science, and technology. Through and beyond fashion, culture is the social production and communication of identities, meanings, knowledge, values, aspirations, memories, attitudes, and understandings of the world—its consideration is fundamental to our ability to live well together.

However, culture was only introduced into policy documents as a fourth dimension of sustainable development in 2001 (UNESCO, 2001). Eleven years later, culture is still only predominantly talked about in Western centric discourse in terms of something to preserve, rather than the means to thrive together. By drawing on cultures, ideas, and belief systems from a wider range of knowledges, cultures can inform our perceptions of ourselves in relation to other life forms. Cultures are a manifestation of beliefs, exemplified through Thich Nhat Hanh's (2017, pp. 14–18) explanation of life as "interbeing", a simple yet profound explanation of life. We do well to acknowledge the contribution that tutors make to cultures and in recognising and expressing cultures as beliefs and values through art and design education.

This research demonstrates a very high level of interest in exploring new approaches to teaching and learning, with relational, participatory pedagogy. However, if art and design education is to really contribute to co-existence, then universities must create whole systems change within the academy and actively contribute to change beyond it. Then the conditions within which our graduates go on to live and work are conducive to the application of sustainability learning. This is a huge challenge for tutors as socially accepted practices and collective patterns of behaviour currently avoid accepting the implications of acting on climate change and other environmental threats. Art and design, including fashion education, have a history of questioning the status quo and upholding that tradition is more important than ever. To achieve this end involves a layered approach to change which, when explored through pedagogy that is congruent with sustainability, offers ways to change how (as well as what) we think and thus understand things. Such a transformation in habits of mind requires us to let go of outmoded ways of thinking, but to do so without perceiving the old and the new paradigm in direct opposition. To detach from outmoded thinking should not involve an aversion to all that it entailed, but rather to learn from what has gone before.

Life on earth involves multiple interactions; we depend on the relationships between elements inside our bodies, our environments, communities, and the wider world. Fashion involves interactions that are cognitive, practical, and emotional in a system that is regulated through professionally accepted knowledge, practical shaping of material into form and expressive responses to perceived context. By changing what is routinely included in fashion education, what is recognised as knowledge and how we think about ourselves in the world in terms of place, time, and relationships, tutors and students can start to think with, rather than about life. Perhaps one of the advantages of teaching fashion at a time of deep planetary and societal upheaval is that it is a discipline that can adapt to change—its practitioners often imagine and make ideas come to life before the theory around the activity is formed. Rather than diminish the rootedness of these practices, a balance between the head, hand, and heart elements of fashion education creates a fundamental understanding of an evolving, not-static set of practices. Signature pedagogies, according to Shulman (2005), nearly always involve active performance by learners, which emphasises the importance of participation and visibility of learners, mediated through the creation of safe spaces where experimentation and co-learning can take place. We are in the midst of unprecedented change, and it is vital that tutors and students learn with and as change, not just about it.

The pedagogies, practices, and ideas presented in this chapter seek to unravel some of the assumptions that no longer serve us, but in so doing, to interweave teaching practices of head, heart, and hands that are congruent with sustainability and art and design (including fashion) education. Fashion tutors face the challenge of measuring up against student, organisational, sector, and societal expectations, each with their own yardsticks. To satisfy and go beyond these expectations involves "developing pedagogies to link ideas, practices, and values under conditions of inherent uncertainty that necessitate not only judgment to act, but also cognizance of the consequences of one's action". (Shulman, 2005, p. 19). Thus fashion education becomes a practice of relational understanding, developed through a discursive, critical consideration of the purpose and intention of the practitioner, as well as of the product, service, system, or organisation it represents. This ongoing, evolutionary process can reveal new perspectives, insights, and ideas for all involved. I, for one, am revelling in the unravelling, but only if, and when I am supporting and supported by others in relationships based on curiosity, integrity, and trust.

References

Bateson, G. (1972). *Steps to an ecology of mind*. University of Chicago Press.

Capra, F., & Luisi, P. (2014). *The systems view of life: A unifying vision*. Cambridge University Press.

Delors, J. (1996). *Learning: The treasure within; report to UNESCO of the International Commission on Education for the Twenty-first Century (highlights)*. https://unesdoc.unesco.org/ark:/48223/pf0000109590

Dunne, A., & Raby, F. (2013). *Speculative everything: Design, fiction and social dreaming*. MIT Press.

Ehrenfeld, J. (2019). *The right way to flourish: Reconnecting to the real world*. Taylor & Francis.

Escrigas, C. (2016, June). A higher calling for higher education. *Great Transition Initiative*. http://www.greattransition.org/publication/a-higher-calling-for-higher-education.

Guattari, F. (2014). *The three ecologies* (I. Pindar and P. Sutton Trans.). Bloomsbury Academic.

Hanh, T. N. (2017). *The art of living*. Rider Publishing Company.

Heron, J., & Reason, P. (2006). The practice of co-operative inquiry: Research 'with' rather than 'on' people. *Handbook of action research: Concise paperback edition*, 144–154.

IPCC. (2022). Climate change 2022: Impacts, adaptation and vulnerability. In H.O. Pörtner, D.C. Roberts, M. Tignor, E.S. Poloczanska, K. Mintenbeck, A. Alegría, M. Craig, S. Langsdorf, S. Löschke, V. Möller, A. Okem, & B. Rama (Eds.), *Contribution of working group II to the sixth assessment report of the intergovernmental panel on climate change* (p. 3056). Cambridge University Press. https://doi.org/10.1017/9781009325844

Jickling, B., & Sterling, S. (2017). *Post-sustainability and environmental education: Remaking education for the future*. Palgrave Macmillan.

King, M. L. (1964). The Noble Peace Prize 1964. *The quest for peace and justice*. paras. 3

Korzybski, A. (1933). *Science and sanity. An introduction to non-aristotelian systems and general semantics* (5th ed.). The International Non-Aristotelian Library Publishers.

Lakoff, G., & Johnson, M. (1980). *Metaphors we live by*. The University of Chicago Press.

Lave, J., & Wenger, E. (1991). *Situated learning: Legitimate peripheral participation*. Cambridge University Press.

Maturana, H. R. (2006). Self-consciousness: How? when? where? *Constructivist Foundations, 1*(3), 91–102.

Murray, P. (2011). *The sustainable self A personal approach to sustainability education*. Routledge.

Naess, A. (2008). *Life's philosophy: Reason and feeling in a deeper world*. University of Georgia Press.

Parker, L., & Harrison, L. (2016). By the fire home. *By the Fire*. https://www.by-the-fire.co.uk

Polanyi, M. (1962). *Personal knowledge: Towards a post-critical philosophy*. Martino Fine Books.

Puig de la Bellacasa, M. (2010). Ethical doings in naturecultures. *Ethics, Place & Environment, 13*(2), 164–165.

Ryan, A., & Tilbury, D. (2013). Uncharted waters: Voyages for education for sustainable Development in the higher education curriculum. *Curriculum Journal, 24*(2), 272–294.

Shulman, L. S. (2004). *The wisdom of practice: Essays on teaching, learning, and learning to Teach*. Jossey-Bass.

Shulman, L. S. (2005). Signature pedagogies in the professions. *Daedalus, 134*(3), 52–59.

Sterling, S. (2009). Ecological intelligence: Viewing the world relationally. In A. Stibbe (Ed.), *The handbook of sustainability literacy: Skills for a changing world* (pp. 77–83). Green Books Ltd.

Sterling, S. (2021). Concern, conception, and consequence: Re-thinking the paradigm of higher Education in Dangerous Times. *Frontiers in Sustainability*.

UNESCO. (2001). *UNESCO universal declaration on cultural diversity*. http://www.unesco.org/new/fileadmin/MULTIMEDIA/HQ/CLT/pdf/5_Cultural_Diversity_EN.pdf

UNESC. (2011). *Learning for the future: Competences in education for sustainable development*. https://unece.org/fileadmin/DAM/env/esd/6thMeetSC/Learning%20for%20the%20Future_%20Competences%20for%20Educators%20in%20ESD/ECE_CEP_AC13_2011_6%20COMPETENCES%20EN.pdf

Varela, F. G., Maturana, H. R., & Uribe, R. (1974). Autopoiesis: The organization of living systems, its characterization and a model. *Bio Systems, 5*(4), 187–196.

Walker, S., & Giard, J. (2013). *The handbook of design for sustainability*. Bloomsbury Press.

Wilkinson, A., & Kupers, R. (2013). Living in the futures: How scenario planning changed corporate strategy. *Harvard Business Review, 91*(5), 118–127.

Williams, D. (2015). Transition to transformation in fashion education for sustainability. In W. L. Filho & L. Brandli (Eds.), *Engaging stakeholders in education for sustainable development at university level* (pp. 217–232). Springer.

Williams, D. (2018). Fashion design as a means to recognize and build communities-in-Place. *She-ji: The Journal of Design, Economics and Innovation, 6*(4), 567–568.

Williams, D. (2019, April 3–5). *Fashion design for sustainability: A framework for participatory practice* [Conference presentation]. Designing Sustainability for All The LeNS World Distributed Conference. https://ualresearchonline.arts.ac.uk/id/eprint/14350/1/LeNSin_full_paper._dilys%20williams.pdf

Williams, D., Puppart, P., Valle-Noronha, J., Crew, J., Stevenson, N., D'Itria, E., Vacca, F., Hasling, K. M., & Riisberg, V. (2021). *Education and research IO2—Framework document for design-LED sustainability education*. Fashion SEEDS.

Williams, D., Stevenson, N., Crew, J., Bonnelame, N., Juospaityte, E., Vacca, F., Colombi, C., D'Itria., Aus, R. Puppart, P., Moora, H., Valle-Noronha, J., Riisberg, V., Ræbild, U., Hasling, K., & Eskholm, L. (2019). *Education and research, The benchmarking report*. Fashion SEEDS.

CHAPTER 13

Strategy for Building a Transformative Relationship Between the Academy and the Social and Business Environment Through Interdisciplinary Work

Gloria Hoyos Bustamante

Introduction

The capabilities for the twenty-first century are different from those in which professors were trained in the twentieth century. The development of students' capabilities for a new world requires training them to face real-life projects, with a positive impact on society. To this end, at the Universidad Nacional de Colombia (UNal)[1] Manizales Headquarters, we

[1] By its acronym in Spanish.

The original version of this chapter was revised: Chapter author's surname has been corrected. The correction to this chapter is available at https://doi.org/10.1007/978-3-031-23152-0_18

G. Hoyos Bustamante (✉)
Universidad Nacional de Colombia, Manizales, Colombia
e-mail: ghbustamante@unal.edu.co

M. J. Lehtonen et al. (eds.), *Design Education Across Disciplines*,
https://doi.org/10.1007/978-3-031-23152-0_13

created a group of interdisciplinary courses as a strategy for relationship building with the local social and business environment through interdisciplinary work and comprehensive training. Methodologically speaking, we embrace ***research through design***: a way of building theory derived from action. This approach differs from ***research for design***, which focuses on design process improvement, and from ***research about design***, which is a theoretical field of study (Frayling, 1994). Thus, ***research through design*** pivots the strategy and nurtures a ***way of knowing*** in action. In this experience, design, used as an articulator of various fields of knowledge, emerges as a catalyst for the transformation in administrative management, inside and outside the academy. We conclude that this strategy implies managing changes in the roles and responsibilities of the academic community, not only for students but also for professors and administrators.

In this chapter, we document our experiences and conclusions as follows. First, we contextualize our institution in order to propose a normative framework, a conceptual framework, and a methodology. Then we present the strategy detailing the nature of every course created and its achievements. Finally, we bring to life the thoughts, difficulties, challenges, and projections of the strategy.

Context

UNal is a public university with a strong national presence. It has headquarters in Bogota, Medellin, Manizales, Palmira, Tumaco, Amazonia, Orinoquia, La Paz, and Caribe. Since 2019, at the School of Architecture and Urbanism (EAU[2]) of the Faculty of Engineering and Architecture (FIA[3]) of the Manizales Headquarters, we implemented a cluster of interdisciplinary courses as part of a strategy to build a transformative relationship between the academy and its external environment.

Since then, we have offered these courses to students from all the programmes of the three faculties[4] of the Manizales Headquarters. Our goal is the comprehensive training of students, linked to real problems, to promote their integration into the business environment and their active participation in local, regional, and national development. Thus, we work towards the consolidation of the virtuous circle of the UNal's mission axes: teaching, research, and extension.

[2] By its acronym in Spanish, Escuela de Arquitectura y Urbanismo.

[3] By its acronym in Spanish, Facultad de Ingeniería y Arquitectura.

[4] Faculty of Engineering and Architecture, Exact and Natural Sciences, and Administration.

The current regulations (Universidad Nacional de Colombia, 2007) establish UNal's student training baseline. The proposed strategy of creating interdisciplinary courses is framed within principles of academic excellence and comprehensive and interdisciplinary training. Aspects of the regulations' spirit are implemented in our interdisciplinary courses: incorporation of new schools of thought and technologies, teaching and learning of critical thinking, systemic thinking, innovation, teamwork, solidarity, and individual and collective responsibility for the well-being of the community. We are guided by the articulation between local and global communication networks. We implemented free-election courses as a means for research, extension, and entrepreneurship tasks, emphasizing the awareness of the social implications of the creation of knowledge.

Developing research skills, teamwork, and applying knowledge in context were enunciated as postgraduate objectives in 2007. By 2021, these are the objectives of our interdisciplinary undergraduate courses. The courses' current potential consists of articulating undergraduate and postgraduate programmes, mainstreaming the different curricular programmes, and addressing real problems of increasing complexity through interdisciplinary work. UNal's regulations state that graduates' comprehensive education and versatility are a means of improving their participation in society and their role in the nation's development.

The conceptual framework is based on the recognition of the design process (Perfes, 2011), and the strengthening of the undergraduate students' acts of creation. In its broadest sense, designing involves at least three moments: ideation, design, and construction. Being aware of this process opens the doors to the self-knowledge of how we design when we design, according to the precepts of Lonergan (1990). For this Canadian philosopher, the self-appropriation of knowledge consists of an elevation of our degree of consciousness that highlights our conscious and intentional operations and leads us to answer these three fundamental questions: what do I do when I know, why is this activity knowledge, and what do I know when I do this activity. The first answer is cognition theory. The second is epistemology. The third is metaphysics in the transcendental sense, an integration of heuristic structures (Lonergan, 1990).

In *multidisciplinarity*, an initial design from a particular discipline is proposed, to which are superimposed final technical, technological, structural, and economic solutions that determine its viability. However, in a project, spatiality, infrastructure, function, cost, and materiality should emerge as defining elements, far from the multidisciplinary linear conception. Interdisciplinary work, understood as the early dialogue between the disciplines convened in a project, allows us to anticipate difficulties and include technical requirements as design parameters. Consequently, we promote scenarios for synergy between undergraduate students of the three faculties as a way for comprehensive *interdisciplinary* training. Thus we optimize resources (make, undo and redo) and it is potentially possible to achieve *transdisciplinary* results: those where we cannot point out from which field of knowledge they emerged.

For Saikaly (2005), *a particular way of knowing*: practical knowledge, is the third area of knowledge. We find it imperative to legitimize this particular way of *knowing through doing* as an emerging approach to research. For Findeli et al. the archetypal task of design and, therefore, of research through design, consists of "improving or maintaining the world's habitability in all its dimensions (physical, psychic, spiritual)" (2008, p. 21). According to this author, those who exercise disciplines related to design stand out for the good organization of knowledge according to the creative act, but their talent can remain tacit. In the interdisciplinary courses created, we apply this *knowing through doing* in undergraduate contextualized training through joint work with enterprises, organizations, and the community. Thus, from the EAU we contribute to this field of research which is under-construction in Colombia, by unveiling this specific knowledge, *a designerly way of knowing*, taking it, in terms of Cross (2007), to a more reflective level and turning it into explicit applied knowledge.

Methodology

We validate "a way of knowing through doing" (Cross, 2007), typical of disciplines related to design, in coherence with the Reference Framework, as an emerging approach. Our approach considers strengthening the students' self-critical knowledge, and the capability for constructing intellectual autonomy, based on their personal way of ideating, projecting, and executing (Períes, 2011), to orchestrate interdisciplinary collaborative work.

We apply what we call an open method: it is in continuous construction, emerging from the synergy between educators and students, with enterprises and communities. Consequently, in order to launch an interdisciplinary course, we begin with a voluntary agreement between the academy and the enterprise. Once mutual trust and a way of working together for the benefit of the course is triggered, we proceed to elaborate on the respective agreements, frameworks, and specific formal requirements for both parties. This way of acting is consistent with *a way of knowing through doing*. Based on these agreements with the local initiative, we then transcend to the national level.

At this point it is worth clarifying that institutions normally proceed in the opposite direction: first, they establish a signed "formal framework", then they elaborate specific agreements to detonate actions at a local level; in our case, from UNal's national level to each headquarters. This approach goes from theory to practice, from "intention" to "application". The drawback of this linear method of acting lies in the fact that the negotiator's administrative deadlines may expire, causing the loss of the achieved inertia, and the dilution of the intention of formalizing a relationship.

From the logic of research through design this process evolves in another way: practice gives rise to theory. Thus, once we trigger the relationship with another institution or company, the expiration of administrative deadlines does not suppose a drawback; the new administrator assumes an ongoing project (the relationship is already established and working) to which only the formalization of the agreements remains to be completed. Under no pretext can the administrative bureaucracy cause the termination of a triggered relationship. Historically, the *Leyes de Indias* (La voz del Derecho, 2015), which includes a set of guidelines to follow for the foundation of new cities in Latin America, was formalized when most of the cities in the new continent had already been founded. How many projects, perfectly written, remain on paper without materializing?

When planning a subsequent version of a course, its subjects are dynamically refactored according to the feedback obtained through the interaction of students, educators, and company tutors, all coming from various fields of knowledge. Ultimately, we assess the resulting exercises produced during the interdisciplinary courses according to their direct relationship with the reality of the problem addressed and the viability of the emerging solution alternatives.

The Strategy

In the context of an interconnected world, creative work of an interdisciplinary nature is seen as a possible route for obtaining transdisciplinary results. Through a cluster of courses, we encourage research as an undergraduate training process with an interdisciplinary postgraduate projection. We anticipate future job opportunities focusing on the real needs of the academy's external environment. We raise our graduate's baseline knowledge, and we aim at improving the institutional processes critical to the professors' and staff's work.

Types of Courses

We offer four types of courses to promote the relationship with the social and business environment through interdisciplinary work: Innovation and Entrepreneurship: from creativity to action (Cátedra Novus[5]), Cursos Empresa FIA, Interdisciplinary Workshop I and II, and a selection of open courses.

Cátedra Novus was created by the Faculty of Mining of the Medellin Headquarters and replicated in the Palmira, Bogota, and Manizales headquarters. In this course, we address challenges from UNal's internal Units and from companies. Faced with these challenges, interdisciplinary teams compete to provide the best solution proposal, expressed as a minimum viable paper prototype. The companies support the course's business model. This aspect makes the course autonomous in economic terms, covering the costs of prototyping processes, laboratory tests, and student training in entrepreneurship, innovation, and soft skills, by experts not on the university's teaching staff. Soft skills are so named as opposed to technical skills. These skills include among others: assertive communication, positive attitude, teamwork with individual responsibilities, organizational leadership, emotional intelligence, adaptation to change, decision-making, and time management. We believe that these life skills, together with technical skills applied in context, enable the achievement of collective objectives and personal fulfilment.

In the **Interdisciplinary Workshop (I and II)** we address the problems of vulnerable communities. Faced with problems proposed by a

[5] Bench named after the methodology used, recognized as Novus.

National Presence UNal Headquarters (first cohort 2020) or by a vulnerable community in the region (second 2021 and third cohort 2022), interdisciplinary teams carry out analysis, diagnosis, and problematization during Workshop I. Afterwards, in Workshop II, teams propose alternative solutions. Students can fulfil their Final Degree Project in a "project participation" modality during Workshop II, with tutors from the student's Basic Academic Unit (UAB), or from a different faculty. This last option represents a quite different opportunity, which had never been done before at our headquarters.

With the **Cursos Empresa FIA** courses, we bring expert, technical, and professional knowledge of business and industry to the university. The first course started in 2019. By the end of 2021 five different courses were on offer. Cursos Empresa FIA courses introduce students to specific technologies, corporate governance trends, markets, human resources, gender equality policies, environmental management, distribution chains, hygiene and safety, production, project management, commercial management, and, if possible, visits to the enterprises' production plants.

With the **Open Courses** we offer an alternative approach to the academy, gradual and informal. These courses do not require registration, attendance control, or evaluation; they do not lead to academic credits and reflect exclusively the student's interest in knowing certain aspects of a particular type of industry. This training, provided by the companies, takes place during weekends or university vacation periods.

Through the Cátedra Novus and the Interdisciplinary Workshop, we provide alternative solutions to real problems in the external environment. For this, we require the support of UNal Research Groups related to the challenges and problems that are the object of the courses during each academic period. This applied research opens an important workspace for doctorate students and graduates, as they explain the use of their training in industry. Meanwhile, the graduate students discover that *there is life after the academy* and that teaching is not the sole profession on the horizon.

Achievements

From the first half of 2019 to the second half of 2021, we had training in soft skills, entrepreneurship, and innovation, subjects absent from current curricula at the UNal, Manizales Headquarters. We strengthened links with the productive sector (local, national, and international companies) and public institutions, through our interdisciplinary courses. We formalized the participation of experts from the companies, who impart knowledge from their experience in the field. We enlist them, without remuneration, to the UNal as Adjunct Professors in the Cursos Empresa FIA. We have exerted a positive impact on the social environment by providing alternative solutions to problems in vulnerable communities, through the Interdisciplinary Workshop I and II.

In addition, with this cluster of courses, we have fostered links within the UNal (intraUNal): between Research Groups, administrative, educators, and students of different programmes of the three faculties of the Manizales Headquarters; with the Headquarters' Student Representation in calls for registration for the courses, scholarship holders, and auxiliary students; with other Headquarters, like Medellín, Bogotá, and Palmira, through Cátedra Novus and Curso Empresa FIA V; and with the Caribbean Headquarters during the first cohort Interdisciplinary Workshop I and II.

We offered a Diploma Course in Dry Construction Project Management (2021), based on the Faculty of Administration, Manizales Headquarters, with professors from the Faculty of Engineering and Architecture of the Manizales Headquarters, the Faculties of Engineering and Arts of the Bogota Headquarters, and the Faculty of Architecture of the Medellin Headquarters. This Diploma Course had one hundred and twelve (112) participants, mostly from Colombia and a few from Ecuador and Panama. We thus bring together academic and theoretical knowledge, the fruits of the postgraduate training of our educators, and the expert, technical knowledge provided by the specialists of the companies in the dry construction sector. It was also intergenerational: the age range of the participants was wide, from young to older adults; multilevel: there were undergraduate students, postgraduate students, technicians from specialized companies, graduates, construction professionals, retired professionals, dry construction system on-site installers, and company managers. It was a very valuable experience given the diversity of the participants and the exchange of knowledge between academia and the productive sector, within a construction system without a great tradition in the country.

Contribution of the Interdisciplinary Courses to the Articulation of UNal's Mission Axes: Teaching, Research, and Extension

From the teaching practice in undergraduate programmes, we provide students with conceptual bases and practical tools to understand and carry out open innovation or entrepreneurship processes (Interdisciplinary Workshop I and II; Cátedra Novus). Thus, students combine academic knowledge, soft skills, and collective and personal interests, contextualized within real social and business needs.

In Cátedra Novus, during the negotiation with challenging organizational, enterprise, or UNal dependencies, we defined the nature of the challenge as extension or research. A "Novus Extension Challenge" points to the design proposal of a product, which, due to its objectual nature, can be materialized. A "Novus Research Challenge" is oriented towards a solution proposal, which can be implemented using explicit routes. Thus, through the extension standpoint, we propose the design of products that solve problems and through the research lens, we support the path to the challenging solution.

In Cátedra Novus we have auxiliary students from postgraduate studies, to whom we provide an economic stimulus for their support work. Utilizing the concept of *gregario* from competitive cycling—a cyclist who in a race or in a lap has the mission of helping the team leader or another more outstanding cyclist of the same team achieve their goals—in each challenge, the respective auxiliary student accompanies and encourages the competing teams during the proposed stages. Each gregario has the function of being the link between the research group that supports the challenge, its director as an expert on behalf of the university, and the competing teams. Through this gregario role we dynamize the process of the challenge, catalyse the knowledge of students and promote the realization of their own learning and research projection. Additionally, since the scope that we define for this course is 'a minimum viable paper prototype', with the resolution of the undergraduate academic exercise, this gives rise to a potential "extension project", desirable as the next stage.

This is the virtuous circle we implement: problems raised by companies or communities are brought to the university through undergraduate teaching, supported by applied research, and projected to the extension for the realization of the alternative solutions proposed. In the case of the Interdisciplinary Workshop I and II, we consider that by consolidating it over time we will give way to participation in internal and external calls, from the local to the national, including some under the figure of the "Solidarity Extension" that exists in the UNal.

Discussion

Our strategy consists of simple and agile relationships, with scarce resources and enormous possibilities. This relationship strategy is usually unattractive to those who only appreciate megaprojects, agreements, and massive contracts. Despite the modesty of the strategy, the bonds forged have provided us with important resources and opportunities: they have opened doors for our students and graduates to internships and jobs. After the bond is consolidated, everything from that point on is net gain: new relationships are unleashed, with opportunities for our undergraduate and graduate students; for our graduates; for permeable directives, and for those of us who are in favour of relevant epistemological events, it allows us to invigorate our passion for teaching.

The implementation of interdisciplinary courses has brought satisfactory reviews from students, educators, administrators, and entrepreneurs. This situation shows that, although historically academia and business have travelled separate paths, it is possible to develop this type of strategy together. Let us remember the African Proverb: "alone, we go fast; together, we go further". Since 2007, the relevance of these courses was already stated in the regulations (Universidad Nacional de Colombia, 2007); however, just adding them to the regulations does not guarantee their implementation. Today's world demands that our students prepare themselves for professional performance in a geographical and historical context; as Barros and Alcadipani (2022, p. 7) remind us: "if all eyes are on the "international" arenas, who would discuss local questions in the language of the locals?". Professional training is a tool to face intellectual challenges and an integrative way to solve real problems. Therefore, disciplinary knowledge cannot become an ideological trench to limit *being-in-the-world*, in Heidegger's sense (1962).

We have identified some difficulties in the following three levels: student, teaching, and administrative. Regarding students, there can't be interdisciplinarity without teamwork. Teamwork is not possible if there is no self-knowledge. Effective and proactive performance is not possible if there is no individual work on self-knowledge, both about the meaning of life and in relation to the other. Self-knowledge makes it possible to manage active participation within a team by putting its skills and deficiencies on the table. Working as a team requires being aware of one's own cultural baggage, pre-obligations, and personal prejudices, in order to delimit and enhance contributions within the collective. Without self-knowledge, the collective is just a group of people assembled in an arbitrary way, each coming from different disciplines, without valuing their own knowledge, and without recognizing the knowledge of others; thus, each one has different horizons, moves in different directions, and lacks mutual respect. This training for life, assumed from Cátedra Novus, seems a requirement not only of the epoch but of the world, of humanity. Currently, we have the technological tools for networking, collaborative, synchronous, or asynchronous, but we still have to work on tuning our inside with our outside. Although training in entrepreneurship, innovation, and soft skills is addressed exclusively from Cátedra Novus up to this point, the practice has shown us that it is a general requirement for the set of interdisciplinary subjects. Each of these courses can lead to an innovation or entrepreneurship project, and in all cases, as an institution of high academic quality, the greatest shortcoming of our students is the training in soft skills.

Regarding the faculty and staff, apathy and uncertainty make things difficult. This is not a static model but a dynamic strategy, in continuous feedback, both in the classroom and in the offices where the processes are legitimized. When challenged by a new situation, generally, the answer is *we do not know how to do that because we have never done it before.* Additionally, the training in skills for interdisciplinary work means getting out of the classroom and leaving comfort zones. This implies, at the same time, having continuous personal training as professors, in order to perform in difficult environments with blurred knowledge boundaries. This situation often requires us to unlearn, relax, and place question marks after those things we have treasured as very small "absolute truths".

Regarding managerial and administrative staff, we require more dynamic processes and the means to overcome the lethargy that exists at big public institutions. In regard to involvement with the environment, it is necessary to answer in real-time and not in academic terms. To consolidate this involvement, we must implement the communication vessels between these three entities, student, professor, and administrator, and among the various dependencies; establish relationships derived from projects and not from egos. The main purpose of the institution should prevail before the *petit* powers of the administrative position.

Also, we require the flexibility of the "collegiate bodies" where the decision-making processes are taking place. Assume that the best negotiation is one in which both parties, and therefore the largest number of people, benefit; that you have to give something to get something in return; that contributions made by interdisciplinary undergraduate teams do not necessarily lead to patents. Ideas are not patented and challenges provide *alternative solutions*, not finished solutions to be implemented. As an academy, we must detach ourselves from a result whose potential paralyzes the relationship due to the legal minutia "in case of ...". From the approach of knowing through doing, the most important thing is joint action; there lies the pedagogical balance for all parties involved.

Rigour, responsibility, and ethics are required by all parties involved in the relationship. We should be aware of the high initial investment of energy, with no promise of a quick return: education is not assessable in immediacy. It is also important to consolidate the strategy so it can overcome administrative delays, changes in individual opinions, and dangerous politicization of internal processes in our public institutions. As a Latin-American public university, we must include a social and solidarity-based economy in our training curricula. In this way, we can tend to balance training in entrepreneurship and innovation, not only from competitiveness but from solidarity.

This strategy for building a relationship between the academy and the social and business environment through interdisciplinary work is transformative because it demands changes in the entire academic community. Interdisciplinary work requires strengthening the team-working training skills from the perspective of the harmonization among humans, technology, institutions, and curricula through projects of collective development as a nation. That is why we intend to position training in soft skills, innovation, and entrepreneurship as the DNA of the strategy.

Being a strategy, not a model, we must advance in the incorporation of new companies, as well as in the creation of other courses and projects of an interdisciplinary nature. The comprehensive, interdisciplinary, and systemic training, concerning the environment, is consistent with the three missional axes as elements belonging to the same structure: "UNAL as a national, cultural and collective project" (Universidad Nacional de Colombia, 2019). We intertwine teaching, research, and extension pragmatically, within a hermeneutic spiral (e.g. Hassemer, 1968): first, undergraduate teaching articulated to postgraduate applied research, and second, in terms of extension, within which the experience in the extension feeds the teaching practice, inspiring again undergraduate research. Thus, as we are trained in the capabilities of contemporary life, we weave the university from within and we articulate with the other headquarters through joint projects with society and the productive sector.

References

Barros, A., & Alcadipani, R. (2022). *Decolonizing journals in management and organizations?* Management Learning, Advance Online Publication. https://doi.org/10.1177/13505076221083204

Cross, N. (2007). *Designerly ways of knowing*. Birkhäuser.

Findeli, A., Brouillet, D., Martin, S., Moineau, C., & Tarrago, R. (2008). Research through design and transdisciplinarity: A tentative contribution to the methodology of design research. Paper presented at the *Focused-Current Design Research Projects and Methods*. Swiss Design Network Symposium, 67–91.

Frayling, C. (1994). Research in art and design. *Royal College of Art Research Papers*, *1*(1), 1993/4.

Hassemer, W. (1968). *Tatbestand und Typus*. Heymann.

Heidegger, M. (1962). *Being and time*. Harper & Row.

La voz del Derecho. (2015). *Diccionario Jurídico: Leyes de indias*. https://lavozdelderecho.com/index.php/actualidad-2/corrup-5/item/3227-diccionario-juridico-leyes-de-indias

Lonergan, B. (1990). *Method in theology*. University of Toronto.

Períes, L. (2011). *Miradas proyectuales*. Nobuko.

Saikaly, F. (2005). Approaches to design research: Towards the designerly way. In *Design System Evolution: 6th European Academy of Design Conference*, Bremen, 29–31 March 2005.

Universidad Nacional de Colombia. (2007). *Acuerdo 033 de 2007 (Acta 11 del 26 de noviembre). "Lineamientos básicos para el proceso de formación de los estudiantes de la Universidad Nacional de Colombia a través de sus programas curriculares". Consejo Superior Universitario.* Régimen Legal. Sistema de información normativa, jurisprudencial y de conceptos. http://www.legal.unal.edu.co/rlunal/home/doc.jsp?d_i=34245

Universidad Nacional de Colombia. (2019). *Plan Global de Desarrollo 2019 – 2021 (PGD 2021). Proyecto cultural y Colectivo de nación. El plan.* http://plei2034.unal.edu.co/fileadmin/Documentos/Capitulo_Componente_Estrategico_-_V3.pdf

CHAPTER 14

Designing Sustainable Designs: Making Designers Future-Ready

Jesvin Puay-Hwa Yeo

Introduction

Design education is closely related to and influenced by global and local technological, sociodemographic, and economic shifts. In this technology-driven, pandemic era, the definition of design has broadened and evolved. It is more outcome-focused, including a focus on transforming businesses, improving quality of life, connecting communities, and even pushing the boundaries of the design industry. For instance, as most countries open up to travel, airports have, or will redesign, stress points and invest in digital wayfinding design or touchless technologies to guide travellers through unfamiliar environments and integrate a range of new behaviours (Ready & Osbaugh, 2020; Ritchie, 2020). Examples include ceiling-integrated lighting systems with biometric sensors to

J. P.-H. Yeo (✉)
School of Art, Design and Media, Nanyang Technological University, Singapore, Singapore
e-mail: jesvinyeo@ntu.edu.sg

M. J. Lehtonen et al. (eds.), *Design Education Across Disciplines*,
https://doi.org/10.1007/978-3-031-23152-0_14

delineate points on the floor for travellers to queue up or direct them to the nearest sanitising station (Ready & Osbaugh, 2020).

The role of design has also transformed and is embraced by non-design industries and people. Design is used in the private sector, such as banking, to promote innovation and strategy to create new value, respond to disruption and capture new markets. Government agencies and citizens also use design to formulate people-oriented public policies and services and deal with challenges ranging from pollution to sustainability to healthcare (DesignSingapore Council, 2021; European Environment Agency, 2022). For example, the neonatal intensive care unit of a major US health system used human-centred design solutions to improve the quality of breastfeeding support for mothers (Crowe et al., 2022) and a pre-market design approach, safe and sustainable by design, was introduced to the European chemical industry to improve product safety and prevent pollution (European Environment Agency, 2022). In the United Kingdom, the average worker also uses design skills, such as creative thinking and problem-solving skills in the work environment to increase productivity (Design Council UK, 2018). Along the same lines, in Singapore, the DesignSingapore Council (2021) has taken action to promote design as a national skill.

The future will demand an innovation-led mindset to drive economic development in the Fourth Industrial Revolution. By 2030, the world will undoubtedly be very different, but many elements will be similar to today, and certain skill sets will still be in demand. According to the World Economy Forum's Future of Jobs 2020, creativity, ingenuity and initiative are among the top five emerging skills. Hence, design education must and will continue to provide the core skillsets of creativity and technical imagination and exploration, design creation, and prototyping, as continuity is just as important as change. On top of the core skillsets, we also need to provide a platform for future designers to innovate and explore and equip them with the ability to self-learn and adapt in order to advance the design applications and discourses and to help deal with the challenges of the twenty-first century, some of which we may have not yet encountered.

With all these changes, a new form of design education must be redesigned to meet the needs of today and allow for the good practice of responsible and sustainable creativity to continue and expand in future. Focused on Singapore, this chapter aims to understand how the industry is changing, how the roles of design education, faculty, and students are changing, and how we can prepare designers for the future.

Emerging Design Roles

With a shift towards a knowledge-based economy model focused on technology and service development, design approaches that were once used in the design sector are now widely accepted in many business and social contexts. In particular, design thinking is used by many organisations to carry out human-centred activities to develop better systems, services, processes, digital interfaces, and more (Ideo, n.d.; Yamazaki, 2014). This change enables design firms to reposition themselves as innovative design consultancies (Brown, 2009). Now, with ever-changing digital technologies and demands for businesses to communicate with stakeholders in real-time, we are seeing the rise of enterprises and service industries, such as banks and hospitals, who have assembled their own in-house design teams to develop content and promotional materials for service and digital design (Muratovski, 2015).

It is for these reasons that the design workforce in Singapore is expected to grow, especially in the non-design sector, with an increase of 60% forecast by the year 2025 according to the 2019/2020 National Design Industry and Manpower Study Summary Report by DesignSingapore Council. The report estimates that for every 1 design job in a design service firm, 2.1 design jobs will be created in non-design firms. As a result, exciting new design roles are emerging, such as Design Researchers and Analysts, who study key industry trends and user experience strategies and gather meaningful metrics to evaluate user interface performance to support the creation of user journeys, wireframes, and prototypes. Next are Product Managers, who are responsible for the product life cycle, from product development to ideation, prototyping, and delivery. In addition, the role of the Experience Designer involves designing and shaping user-centric products and experiences. Finally, Strategists, aka Business Designers, will strategically plan activities that will enhance an organisation's brand, products, and services (DesignSingapore Council, 2019b).

Furthermore, as technology, design, and business converge to build innovation and explore future possibilities, interesting new roles have also been identified globally in the expanding field of design. Including AR Journey Builders, who work with talented artists and engineers to create augmented reality (AR) experiences for consumers; Digital Tailors who advise customers on fabric and clothing choices and styling while also recommending the latest fashion trends and upcoming events; Vision Practitioners who question assumptions and inspire people; Business Model Designers who develop actionable operational strategies; Language Manipulators who manage communications; Scenario Writers who create storylines and plots (Pring et al., 2017; Xiang & Liu, 2018).

So, while the design industry continues to define jobs, design education needs to broaden its definitions of what qualifies as an occupation. As all forms of work will become vulnerable in a globalised and challenging environment, design education must provide a platform for future designers to focus more on being at the core of ideas and innovation that support their learning and development.

Interdisciplinary Skills

In a knowledge-based economy, designers with interdisciplinary skills will be very much required in the future. In fact, the role of the designer has gone beyond simply brainstorming a brand, product, and/or service to create strategic advantages to transform systems, experiences, and organisations (Singapore Design Council, 2019a). In addition, designers are invited to participate in other businesses to identify larger challenges and propose more meaningful solutions that cannot be solved by designers or executives alone (Sanders & Stappers, 2012). Designers also need to learn from, and collaborate with, like-minded statisticians and others to develop new and appropriate methods for this increasingly complex world (Norman, 2010). For these reasons, future design education must require strategic alliances with industry and even other educational institutions to engage, educate, and foster interdisciplinary learning for future designers.

In Singapore, design education has moved beyond the teaching of practical design skills to provide students with exploratory and interdisciplinary skills that enable them to ask relevant design questions and examine research pertinent to their personal formative practice as creative individuals (Yeo, 2022). For example, design students in the NTU University Scholar Program, collaborate with non-design students in the core course (Fieldwork and Documentation: Topics in Sustainability) to hone their discipline-specific skills and develop a multidisciplinary approach to the planning and execution of a sustainability research project at the nexus of economy, environment, and societal space. Project outcomes are delivered in exhibition displays and research reports. The course aims to cultivate students' ability and interest in promoting mitigation strategies that improve their livelihoods and those of their fellow global citizens. Within the formal curriculum, many design institutions globally and locally have provided opportunities for transdisciplinary learning by allowing design students to gain exposure to non-design subjects and bordering skillsets (Abu Bakar, 2020; Costantino, 2018; DesignSingapore Council, 2019a; Yeo, 2022).

In addition, to nurture new types of designers and non-designers who can work across disciplines and understand design, people, business, and technology, design education must move from being a mere school of art and architecture to one that integrates science, business, and engineering (Yeo & Teo, 2020). For instance, the National University of Singapore recently took a step in this direction by merging the School of Design and Environment and the School of Engineering to form the College of Design and Engineering (Ang, 2021). Students are allowed to take courses in both schools to develop competencies in different areas of study.

In industry, companies and organisations are also beginning to use design knowledge to find unique and creative ways to enhance their guiding principles, brand image, and build stronger customer relationships. For example, the healthcare industry uses design experiences to increase patient engagement, strengthen the doctor-patient relationship, and even seek to change patient behaviour (Caposino, n.d.). This change creates a new context for the design process, with new forms of design emerging from new environments, new tools, new situations, and new technologies (Friedman, 2019). This indicates that the discipline of design can aid advancing economies around the world. However, design education must also reform its curriculum besides continuing to provide core creative and technical skills. This would not only include interdisciplinary learning but also deepens students' worldviews by creating shared learning experiences in areas such as science and technology, health, and wellness. Designers of the future will excel in professional design and possess skills that transcend disciplinary norms.

Social and Technological Shifts

More than two decades ago, Archer (1995) argued that as our awareness and attention to environmental and social issues increases, learning collaborative skills and knowledge of the humanities (such as theology, philosophy, ethics, and literature) is essential. These skills will enable designers to build the meaning of human expressions and categorise, map, and observe the origins of concepts, people, things, and occurrences. Communication design, logos, websites, animations, and even architectural productions have all come a long way since the dawn of the digital revolution. Today, technology plays a vital role in creating meaningful and expressive digital works that convey corporate and policy messages and raise awareness of global challenges such as climate adaptation, inequality, and health.

Design education can help strengthen societal values by harnessing the power of digital and design technology. Future designers should be encouraged to communicate social values through design, reflect on human experience and make value judgments (e.g. Frascara & Noël, 2012; Sanders & Stappers, 2012), collaborate with others on specific topics to identify target audiences' needs or to expand their understanding of norms, practices, and values on social issues (e.g. Caruso & Frankel, 2010; Chick, 2012). As Hvid (2010) argues, "… to address large-scale issues such as climate change, climate adaptation, inequality, education for all people and global health, we simply have to stop doing one, one, one. We have to be able to scale, and we have to be able to combine."

Along with the improvement of digital design technology, design software is readily available on laptops, mobile phones, and Adobe Creative Cloud. Graphic designers, interactive media designers, and digital product designers can now work from their respective creative spaces anywhere in the world, and companies have started hiring freelance designers remotely. This trend has changed the way designers work, and they now must be more independent, resourceful, and proactive, with solid time-management skills.

A new framework is needed that will allow all designers to work across sectors to develop the technologies they need to support business-related activities, solve the eternal problems of poverty, unemployment, and inequality, and allow them to create a new world order in which humanity is placed above profit and power. In addition, perceptions of work are changing due to exposure to technology, social justice, and networks, as designers can now choose to freelance and work on short-term projects rather than full-time jobs.

Making Designers Future-Ready

Design education of the future must be an inspiring space for design, culture, and technology. Ideas and practices, methods and materials, ideologies and philosophies can coexist and be shared among designers, industries, and communities. The role of design education, design faculty, and students must be redefined to prepare for the future (Table 14.1). In this context, the question is how to prepare design students for the transition to this near future?

The Role of Design Education

Many design institutions find themselves at the crossroads between traditional, results-focused design and disciplinary and extended field design. To keep up with the changing world, design education must continue to create and develop a systematic body of knowledge for design, strengthen links with industry, and develop principles and theories that help students understand the complexities of current issues. Design education should collaborate with other disciplines to foster a culture of collaboration between design, science, and technology.

With the development of new technologies, different industry business models, new cultural differences, and concerns about ethical issues, design education must focus on improving students' digital skills and enhancing their technical understanding in the emerging fields, starting with learning data visualisation, computational intelligence, and natural language technologies, as well as developing their soft skills to address issues regarding ethics, fairness, sustainability, and environmental protection.

Furthermore, to ensure stable and expanded employability in the design disciplines, design education must be more transparent than ever. The shifting and changing nature of traditional design fields must be made known to faculty members and students to avoid directing students down blind alleys. Students must also be exposed to dramatic changes in the world's resources and economies, prepare for emerging roles in design and non-design sectors, as well as understanding that design in an extended field may be as crucial to the design discipline itself as it is to society and designers.

Table 14.1 Making designers future-ready

	Emerging design roles	*Interdisciplinary skills*	*Social and technological shifts*
Design education	Design education needs to broaden the definition of occupations, expose students to the dramatic changes in the world's resources and economies, and prepare them for emerging roles in design and non-design fields	Design education must provide students with interdisciplinary learning, exposure to non-design disciplines and bordering skillsets, and collaboration with other disciplines to foster a culture of collaboration between design, science, and technology	Design education can help strengthen students' digital skills, accelerate their understanding of technology, and develop soft skills to address ethics, equality, sustainability, and environmental protection issues
Design faculty	Design faculty should reintegrate into the industry, learn new norms of design and operations, and understand emerging design roles and skills	Design faculty must support students' interdisciplinary learning and experience by providing the tools, direction, and resources for exploration, integration, ideation, and expression	Design faculty must upskill and become digitally literate to provide capabilities beyond emerging technologies to assist students in learning and applying knowledge in real-world situations
Design students	Design students must be able to develop new ways of communicating business, awareness, and education, deliver new systems for healthcare or innovation operations and be flexible and compassionate to meet the needs of emerging design roles	Design students must create designs beyond academic requirements, disciplines, and reasons. They will co-design and apply techniques and strategies they have learned to their daily creations and activities	Design students must understand their advantages relative to robotics and automation technologies, have good judgment and decision-making skills to deal with the subjective aspects of data analysis and design solutions, and convey meaning and significance through design exploration

The Role of Design Faculty

As the world progresses, the role of the design faculty should go beyond teaching, disseminating, and imparting basic or applied knowledge to students, and assisting students in learning and applying knowledge in practical situations. For example, in a world that relies on ever-evolving technologies, the demand for design faculty with digital skills to match is also increasing. We have heard of STEM, but how many of us know about Social, Mobile, Analytics, and Cloud (SMAC)? These digital buzzwords are here to stay whether we like it or not, so faculty members need to be digitally literate to provide capabilities beyond emerging technologies such as artificial intelligence, the Internet of Things, and data science. Design faculty need to respond to the multitude of challenges and possibilities that come with it. They must be able to adapt to change and conceptualise multiple complex ideas simultaneously.

In more open teaching and learning environments, the role of design faculty has shifted from instructor to facilitator. Students will take on the role of lead designer throughout the process of exploring and experimenting with their designs and will play an important role in the development of knowledge, ideas, and concepts. In generating design outputs and interdisciplinary insights, design faculty must have cognitive flexibility to assess and support student development and experience by providing tools, direction, and resources for exploration, integration, ideation, and expression. To push students beyond their core disciplines, interests, passions, and creativity, design faculty have to be facilitators that wear many hats. They will guide students through real-life practical exercises, mentor students to tap into multiple senses and integrate theory with practice, provide support and scaffolding for students to be ethical and critical about their own performance and that of others, and offer a clean slate for students to design for complex situations and speculate on the future.

In addition, for a more sustainable system, the role of teaching should incorporate the role of learning. Design faculty must engage in lifelong learning and continuous upgrading to impart knowledge relevant to the future economy (Steinert, 2017) and provide the essential knowledge students will need in the future. Along with regular sabbaticals and fellowships that are more focused on research, design faculty should reintegrate themselves into the industry, through such things as temporary placements as designers, to learn about new norms of design and operations, as well as to be *au fait* with emerging design roles. They will then be better prepared to craft a curriculum relevant to industry needs, help students create meaningful design outcomes that are practical and desirable, and equip them with the right balance of existing and emerging skills.

The Role of Design Students

In traditional classrooms, student learning may be passive, and some students still pursue an education with economic rewards in mind (Cuellar et al., 2022). In the modern classroom, design students should explore ways through which they can find meaning and joy in their work. They should be able to apply their skills, ask questions and understand their educational and personal needs. So, the role of students will change from passive learning to active acquisition of new knowledge that will help them grow into design thinkers who can collaborate with people from other disciplines and deliver novel and valuable solutions to various industries.

In addition to ensuring that what they learn is understood and applied in an academic setting, the role of students will be full-time and 24/7 as they will continue to create designs beyond academic requirements, disciplines, and reasons. They will apply the techniques and strategies they have learned to their daily creations and activities. A good example from Singapore is DOT, a group of graphic and interactive students who explore the Death of Typography outside the classroom through co-creation with both designers and non-designers, typography activities, and self-exploration in order to address design and contextual values.

The role of design students will also shift from learning how to design products and graphic design to developing new ways of communicating business objectives, awareness, and education, delivering new systems for healthcare systems, and even new tools for non-designers to express themselves creatively. Future designers should be agile, versatile, and empathetic to meet the needs of emerging design roles. For example, as the demand for healthcare jobs increases, students must develop their emotional and social intelligence. These qualities are absolutely essential for them to collaborate and communicate effectively with others.

Conclusion

New technologies and challenges will emerge and require new and innovative tools and approaches for designing. Design education needs to offer new curricula for students to learn experiential and interdisciplinary skills that allow students to collaborate and perform effectively in the design process of idea generation, designing, and production. Students will be encouraged to actively shape their learning and potential and grow as design thinkers by acquiring marketing, analytical thinking, technology, and communication skills and exploring design knowledge boundaries through responsive and collaborative design creation. Consequently, by broadening their intellectual creativity and innovation capabilities, future designers can expand their reach to support different industries that address economic and social challenges.

Design education also needs to provide design faculty with opportunities and alternative ways to advance their digital knowledge and abilities to design exciting project briefs and build scaffolding to support the creativity of students at all levels. Along the same lines, design faculty must foster an open and encouraging culture to broaden intellectual creativity and innovation in design practices.

In conclusion, the economy of the future is complex and will present more challenging issues than ever before. Design education must provide our current students with knowledge and skills that transcend disciplines to create better and sustainable designs for the future. Future design education will:

- assist students to master marketing, research approaches, and data analysis skills to make sound judgements and decisions.
- enable students to advance their digital skillset through learning digital collaboration tools and data visualisation to offer design strategies for digital platforms.
- encourage students to explore the boundaries of design knowledge through responsive and collaborative design creation and cross-collaboration in different industries to establish cognitive flexibility.
- motivate students to investigate social and world issues through uniquely human capabilities to developing emotional and social intelligence.

- encourage students to apply acquired knowledge and skills to find innovative solutions to multidisciplinary problems throughout their professional lives.
- develop students' abilities to acquire critical knowledge and skills that link pertinently to contextual and future work situations.

With these experiences and skills, students will have the ability to use intuition to understand our environment, communicate ideas and solutions that are relatable to others, and be leaders at the forefront of innovation through insightful and relevant interdisciplinary practices. They will live and breathe design and use it to transform businesses and work together to create better sustainable design and living environments.

References

Abu Bakar, N. H. (2020, December 11). Developing a new norm of thinking in design education. *Global Education News*. https://qs-gen.com/developing-a-new-norm-of-thinking-in-design-education/

Ang, H. M. (2021, August 27). NUS Engineering faculty and School of Design and Environment to form new college. *CNA*. https://www.channelnewsasia.com/singapore/nus-engineering-faculty-and-school-design-and-environment-form-new-college-2138861

Archer, B. (1995). The nature of research. *Co-Design, 2*, 6–13.

Brown, T. (2009). *Change by design: How design thinking transforms organisations and inspires innovation*. HarperCollins.

Caposino, M. (n.d.). *The role of graphic designers in the healthcare industry*. FuseLab Creative UI/UX Design Agency. Retrieved May 20, 2019, from https://fuselabcreative.com/the-role-of-graphic-designers-in-the-healthcare-industry/

Caruso, C., & Frankel, L. (2010, July 7–9). *Everyday people: Enabling user expertise in socially responsible design* [Paper presentation]. Design Research Society: Design and Complexity.

Chick, A. (2012). Design for social innovation: Emerging principles and approaches. *Iridescent: Icograda Journal of Design Research, 2*(1), 52–61.

Costantino, T. (2018). STEAM by another name: Transdisciplinary practise in art and design education. *Arts Education Policy Review, 119*(2), 100–106.

Crowe, B., Gaulton, J. S., Minor, N., Asch, D. A., Eyet, J., Rainosek, E., Flint, K., Joo, J., Chambers, C., Bright, S., Yang, J. J., Beyt, G., Pierce, R., & Moses, J. M. (2022). To improve quality, leverage design. *BMJ Quality and Safety, 31*(1), 70–74.

Cuellar, M. G., Garcia, A. B., & Saichaie, K. (2022, January 13). Beyond social mobility, college students value giving back to society. *The Conversation*. https://theconversation.com/beyond-social-mobility-college-students-value-giving-back-to-society-171279

Design Council UK. (2018). *Designing a future economy. Developing design skills for productivity and innovation*. https://www.designcouncil.org.uk/sites/default/files/asset/document/Designing_a_future_economy18.pdf

DesignSingapore Council. (2019a). *Charting the future of design education*. A Report by the Design Education Review Committee. https://www.designsingapore.org/resources/design-education-review-committee-report.html

DesignSingapore Council. (2019b). *National design industry & manpower study (NDIMS) 2019b/2020 summary report*. https://www.designsingapore.org/resources/ndims.html

DesignSingapore Council. (2021, November 24). *Design as a national skillset for Singapore* [Presentation]. Nanyang Technological University.

European Environment Agency. (2022, February 7). *Designing safe and sustainable products requires a new approach for chemicals* (Briefing no. 29/2020). https://www.eea.europa.eu/publications/designing-safe-and-sustainable-products-1

Frascara, J., & Noël, G. (2012). What's missing in design education today? *Literature and Language Journals: Visible Language, 46*(1/2), 36–53.

Friedman, K. (2019, October 3). *Design education today: Challenges, opportunities, failures* [Lecture presentation]. Chatterjee Global Lecture Series, Tongji University. https://www.academia.edu/40519668/Friedman_2019_Design_Education_Today_Challenges_Opportunities_Failures

Hvid, K. (2010). *INDEX: Design to improve life*. The Index Project. https://issuu.com/index/docs/index_designtoimprovelife

Ideo. (n.d.). *Journey to mastery*. Retrieved December 12, 2021, from https://designthinking.ideo.com/journey-to-mastery

Muratovski, G. (2015). Paradigm shift: Report on the new role of design in business and society. *She Ji, 1*(2), 118–139.

Norman, D. (2010). Why design education must change. *Core77 Design Magazine and Resource*, 1–6. http://www.core77.com/blog/columns/why_design_education_must_change_17993.asp

Pring, B., Brown, R. H., Davis, E., Bahl, M., & Cook, M. (2017). *21 jobs of the future: A guide to getting—and staying—employed for the next 10 Years (November)*. Cognizant and Center for the Future of Work. https://www.cognizant.com/us/en/whitepapers/documents/21-jobs-of-the-future-a-guide-to-getting-and-staying-employed-over-the-next-10-years-codex3049.pdf

Ready, B., & Osbaugh, T. (2020, May 11). *It's time to rethink signage and wayfinding for transportation hubs*. Gensler. https://www.gensler.com/blog/time-to-rethink-signage-and-wayfinding-for-transportation

Ritchie, K. (2020, July 10). *How airport design can restore passenger confidence*. Gensler. https://www.gensler.com/blog/how-airport-design-can-restore-passenger-confidence

Sanders, E. B. N., & Stappers, P. J. (2012). *Convivial toolbox: Generative research for the front end of design*. BIS Publishers.

Steinert, Y. (2017). Faculty development: From program design and implementation to scholarship. *GMS Journal for Medical Education*, *34*(4), Doc49.

World Economy Forum. (2020, October). *The future of jobs report 2020*. https://www3.weforum.org/docs/WEF_Future_of_Jobs_2020.pdf

Xiang, C., & Liu, W. (2018). Emerging roles in design expansion: 'The third group' involved in design. In *MATEC web of conferences 176, 02021* (pp. 1–6). https://www.matec-conferences.org/articles/matecconf/pdf/2018/35/matecconf_ifid2018_02021.pdf

Yamazaki, K. (2014). Design thinking and human-centered design—Solution-based approaches to innovation and problem-solving in social environment. *Social Value Design, Special Issue: Contributing to Social Value Innovations, 8(*3). https://www.nec.com/en/global/techrep/journal/popular/2019first/10.html

Yeo, J. P. (2022, January 8). *Design education in context: Singapore* [Paper presentation]. Asian Design Education Conference.

Yeo, J. P., & Teo, C. (2020). Exploring transdisciplinary learning and life-long training in visual communication design education. In S. Boess, M. Cheung, & R. Cain (Eds.), *Proceedings of DRS2020 international conference, Vol. 4: Education* (pp. 1817–1827). Design Research Society. https://dl.designresearchsociety.org/conference-volumes/27/

CHAPTER 15

A Meditative Design Development Framework Toward Post Human-Centered Transformation of Pedagogical Processes

Tokushu Inamura

Introduction

The learner is facing increasing uncertainty from environmental, social, and economic shifts, including everything from escalating natural disasters to pandemics, conflict, and societal tensions (Karalis & Raikou, 2020; Schwartzman, 2020). It is of importance and urgency then, to empower learners in such a way that they can imagine and become actors to make the diverse impacts required.

The field of design studies is expanding as it moves toward solving global, social, and environmental issues. The expansion of the field continues as it becomes further entangled in wicked problems that are planetary in scope. The planetary limits presented by Rockstrom et al.

T. Inamura (✉)
Kyushu University, Fukuoka, Japan
e-mail: inamura@design.kyushu-u.ac.jp

M. J. Lehtonen et al. (eds.), *Design Education Across Disciplines*,
https://doi.org/10.1007/978-3-031-23152-0_15

(2009) and the limitations of human-centered design as argued by Thomas et al. (2017). There are also organizational bodies such as DESIS, spearheaded by Ezio Manzini (Escobar, 2018). The Global Goals Jam, in which the applicant participates as a practitioner in collaboration with the United Nations Development Organization, also considers the planetary limits presented by Rockstrom et al. (2009), and has been adapted into a design tool for doughnut economics, among others (van Woerde, 2018). Though promising fragments are emerging, foundational methodologies for post human-centered design have not yet been established.

Meditative Methods and Approaching the Deep

Though no single change will likely cause such transition, the working of the mind and the inner self is a crucial and persistent factor. Here, Transition Design framework is important as it highlights mindset as a key element (Escobar, 2018). Meditation and associated practices of awareness are promising in this respect. Though ancient in origin, meditative practices including mindfulness meditation and yoga have displayed efficacy in multiple clinical applications (Chimiklis et al., 2018; Goldberg et al., 2018). According to Kabat-Zinn's (2003) definition, mindfulness is "an awareness that is manifested through deliberate attention without evaluation to the moment-to-moment experience that unfolds in the present moment". The research in the applications of mindfulness in the current sense is highly influenced by earlier efforts by Kabat-Zinn (1982), who reported its application in the relief of chronic pain and its development as Mindfulness-Based Stress Reduction (MBSR), which was later called the MBSR program. Since then, clinical trials have reported evidence of various effects, including stress reduction (Levy et al., 2012).

Meditation is not only effective for relaxation and therapy, but also in education and creativity. In relation to education, improvements in cognitive function and emotional regulation in school settings have been reported (Sibinga et al., 2016; Zenner et al., 2014). Further, laboratory-based research as well as meta-analysis of the literature are showing that there are significant causal relationships between creativity and meditation (Lebuda et al., 2016).

In addition, reviews and meta-analyses have revealed the effects of meditation on empathy for others, including non-humans, in the design process (Luberto et al., 2018). This is an important area of research for design pedagogy as it takes on the challenge of post-anthropocentric "wicked problems" that require an increasingly high degree of empathy and creativity to design beyond human concerns.

According to Sanders and Stappers (2014), design is shifting from an era of design with to design by, from an era of users to an era of participants, from designer-led user involvement, called co-design, to collective dreaming, where the potential aspirations of diverse participants are realized together. The shift is from designer-led user participation, called co-design, to collective dreaming, in which the latent aspirations of diverse participants are realized together (Sanders & Stappers, 2014). Visser et al. (2005) have conceptualized the importance of generative sessions in co-design to address the tacit and latent aspects of participants. However, there is much room for further research on this tacit and latent knowledge, which can be directly approached through meditation and mindfulness. As such, in this Chapter 1 will touch upon prior work in design, engineering, and innovation from the perspective of exploratory frameworks on meditative praxis in interdisciplinary education. Moreover, I will also cover some implications meditative praxis might have for instructors and learners.

Meditative Research and Designing

Given that there is a paucity of studies on meditative methods especially in relation to design, there are ample opportunities for conducting impactful research in this domain. For instance, Rojas et al. (2015, 2016) has conducted a thorough narrative review of work in mindfulness and design, as well as pointing out the importance of mindfulness in design. Pahl (2005, p. 10) has conducted preliminary clinical work with engineering design students who were novices at Diamond Way Buddhist meditation, with results suggesting that students "feel calmer under stress and enhance lucidity or awareness during moments of decision-making". Further, Niedderer et al. (2020) have developed a "mindful interdisciplinary co-design methodology" and compiled approaches to incorporating mindfulness into design for co-design with dementia patients.

One of the most recent theoretical inroads to mindfulness in design has been proposed by Andrahennadi (2019) who has developed a Mindfulness-Based Design Practice (MBDP). It is based in Buddhist practice and directly takes learners through Buddhist meditation including water meditation. Similarly, Akama (2012, 2014) has published extensively specifically from a Zen Buddhist perspective as well as illustrated the implications mindfulness has in design practice; this standpoint can be viewed as a Research through Design (RtD) approach and one that is based on a continuous practice rather than pragmatic applications of meditative tools. Simpson-Little and Long (2010) have conducted various meditative sessions, in which guided visualization sessions were included. A session was reported where new products were designed, through having participants explore a famous designer's house in their imagination, followed by imagining that they were a lamp made by the said designer, and to then draw it. This is a rare example of where meditation has been used to ideate, and this method has many overlaps with what will be covered later on in this chapter. It can be said that there is currently a certain lack of overall strategy toward developing a secular Meditative Design and associated pedagogy.

Types of Meditation

In considering the research potential of effective meditative methods in design, it is necessary to cover the significant types. The two major meditation categories are focused attention (FA) and open monitoring (OM). FA is a meditation method that focuses attention on a single object, while OM is an observation of an individual's perceived thoughts, sensations, etc., without fixating on any object, i.e., the object of attention is flexible and unrestricted (Colzato et al., 2012). Both FA/OM have been reported to have a causal relationship with creativity, but an analysis found that OM meditation was particularly effective and likely to improve creativity (Lebuda et al., 2016). Such evidence-based meditation research is highly promising, yet the limitations of current research approaches suggest that there is a paucity of opportunities to apply research findings to practice (Henriksen et al., 2020).

Meditative Processes as Part of Design Research and Embodied Praxis

Given the dearth of research, design as synthetic transdisciplinary praxes should then take the lack of applied research as a challenge, and engage in teaching and learning from meditative RtD outputs so as to enrich the body of work on the theory of mind, as well as develop novel methodologies which will positively impact the Anthropocene. The author has been developing personal praxis to this end, having engaged in meditative processes through training in multiple martial arts, including influences from Zen, Yoga, Taichi, Qigong, and bioginástica. Meditation formats have formed an important part of an embodied design and engineering research program. Drawing from such experiences, in combination to academic learnings from meditative literature, the following section will outline a sample of meditative praxis undertaken.

Meditative Design Praxis

Meditative activities, including meditation, have been implemented in multiple educational settings including long-term RtD projects, studios, multi-day workshops, and seminars between 2018 and 2021. Cumulatively over one hundred students were guided in meditative exercises including undergraduate students. The majority of the students conducted the exercise online using Zoom/Teams video conferencing and Miro whiteboards in modules in Media, Art and Design Thinking, offered to undergraduate students in the 2nd year onward in the School of Design and School of Interdisciplinary Science and innovation. Student impressions (using the meditative ideation method detailed in further sections) were collected, however, for the purpose of this chapter, the most interesting examples and feedback are from face-to-face classes, since they were comprised of small teams of less than five.

Kyoto Kamo River Meditative Ideation Sessions

The first example was completed with four students in the 3rd year Information Science and Engineering Bachelor's degree course, conducted in 2021. They were to develop post-human-centered solutions looking at life on the Kamo river. Meditative ideation sessions were conducted where guided meditations were conducted over Zoom. Students had previously conducted field work on the Kamo river that flows through Kyoto, the former capital of Japan. All students were familiar with the river and were to design interventions that included digital fabricated outputs. One session of meditative ideation was conducted with the following feedback.

> *One of the things that left a deep impression on me during the mindfulness session on the Kamo river this time was that I could imagine many bad aspects of the river (environmental pollution, especially littering), but could not imagine many good aspects of the river…I thought that by doing this technique and writing it down before the fieldwork, I would be able to get new ideas for design and differences before and after the fieldwork.* (Student 1)

> *Meditation made me think more about what I really envisioned and what I wanted to do by thinking about my ideal world and getting it on paper. However, I also felt that it was difficult to come up with a design that matched the ideal and the reality.* (Student 2)

> *Although I was a little sleepy at some parts, I was able to think about people, the Kamogawa River, ducks, and other animals and plants from a bird's eye view by continuing to meditate. I don't think I would have been able to see this perspective if I had proceeded with the project as usual, so I felt that experiencing this design technique broadened my way of thinking.* (Student 3)

Students were able to explore ideas that they considered beyond what they might have imagined. In particular the students had led their attention to the river's condition and mallards who live near it. The guidance provided to the students urged them to both look out for human and other forms of life as well as the name of the river itself (鴨川 literally means mallard river).

Naka River: Designing with the River

In another example from 2019, three Master's level students, one from product design, and two from interior architecture majors (all were exchange students, German and Italian nationals) conducted a design for the Naka river that flows through Fukuoka city in Japan. After field work along the river, meditative ideation was conducted on campus in person.

Figure 15.1 illustrates a lively scene of community with urban infrastructure, people as well as butterflies and trees. Seen from a more realistic natural scene, Fig. 15.2 illustrates how one of the participants was, through the drawing, making parallels between the Naka river and a river located in their home country. Further, Fig. 15.3 shows a more imaginative and conceptual vision of buildings with slides going into the river.

Such variation shown across outputs A, B, and C illustrates the divergence of visions that can occur. The generative nature of the exercise captures the latent thoughts, which are of a different quality from those of frantic ideation sessions whose aim is to produce multiple ideas within a short period of time. These visions were then taken further coupled with explorations of the river on stand up paddle boards, plus further fieldwork, to develop a fully fleshed out proposal to the local government (Nakagawa Project, 2022).

The proposals were then taken to local government environmental departments to explain the future vision. In addition, the vision was shared with invited members of the Nakagawa Future Conference that included public and private organizations as well as local citizens and researchers. The concept was well received and used as a vision to apply for government funding for sustainable riverfront management, which was subsequently approved. A practising architect who is also a researcher from the author's institution was onboarded to develop a bamboo architecture proposal developed with other architectural students, which is

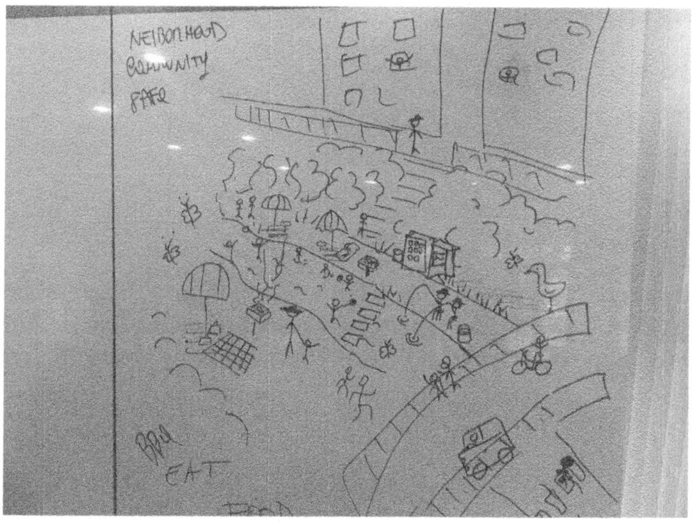

Fig. 15.1 Meditative Ideation output A

Fig. 15.2 Meditative Ideation output B

currently under negotiation to begin construction in September 2022, pending approval. This is a concrete example of meditative design going from classroom to societal implementation, demonstrating the power of the approach.

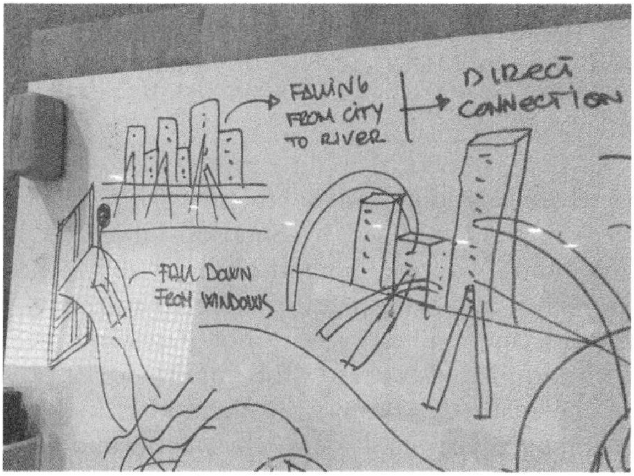

Fig. 15.3 Meditative Ideation output C

Meditative Design Methods

Meditative Ideation (MI)

The sequence for the method is outlined in the following steps:

Preparations: Ready a writing implement such as pen and paper, whiteboard and marker, etc.

1. Assume a seated position, remembering to sit with good posture.
2. FA meditation breath awareness, with eyes closing after long breaths, followed by imagining a scene of importance, e.g., a riverside, or art gallery.
3. The participant is asked to explore the space in which they are guided and encouraged to look around.
4. As the participant continues to look and explore, questions about who or what is nearby is asked.
5. Following this, the participant is asked to imagine a change in the space/place in a desirable way, to envision a design.
6. The participant is gently to return back to the real space they are in.
7. Following this the participant returns to focusing on their breathing.
8. Attention is back to the room and eyes are open.

9. The participant is then asked to draw what they saw during the session.
10. The educator and participant/s discuss the drawings and their significance.

The "*Guided meditation and creative visualization for Product Designers and Engineers*" method outlined by Simpson-Little and Long (2010) has much in common with the method outlined above. Key differences between their method and MI is that the latter has a stronger post human-centered emphasis, and applications are not necessarily tied to product design. Furthermore, MI also stresses the importance of playing relaxing music during the method. Overall, the method is highly divergent in terms of developing unexpected ideas when compared to more "traditional" ideation methods. In addition, ideas born of MI are not necessarily developed further, but used as inspiration.

Meditative Persona (MP)

1. Assume a seated position, remembering to sit with good posture.
2. FA meditation breathing awareness, with eyes closing after long breaths, followed by imagining the person (incl. non-human) that the design proposal will influence.
3. The participant is asked to imagine the proposal impacting the person.
4. The reaction from the person is noted.
5. The participant thanks the person for cooperating.
6. The participant is gently to return back to the real space they are in.
7. Following this the participant is again back to focusing on breathing.
8. Attention is back to the room and eyes are open.
9. The participant is then asked to describe what they saw during the session.
10. The educator and participant/s discuss the drawings and their significance.

This method has been devised as a means to create vividly visualized and interactive persona, especially when an idea that has been developed can be tested to imagine conversations that might arise with unexpected stakeholders. Or alternatively it can be used to imagine stakeholders in their daily life where scenarios of their life might come to mind in an experiential form. In the author's own experience of conducting the technique, the author became a crow, surveying the urban landscape (see also Akama & Light, 2015).

Framework

The previous section outlined an example of meditative design methods that can be developed and used in design pedagogy, however, as discussed earlier, there are many types of meditation, and possible variations are as diverse as the co-designing practitioners. Given the potential of meditative methods, how could further opportunities be explored? A sensible hypothesis is that some frameworks would allow for strategic exploration and development. The following section explains the rationale behind using the double diamond (DD) as a base for a Meditative Design Development Framework (MDDF).

DD was developed by the British Design Council, although it has its traceable roots in the cyberneticist Banathy's work in the guided evolution of society (de la Rosa et al., 2021). The strength of this way of capturing the design process is that it reflects the temporal aspect and the divergent and convergent modes that structure design processes in general. Further, the divergent and convergent modes in turn are derived from the work of Guilford (1956; Draaijer, 2009). The alignment of the work to connect the work in design to the science of the mind is important. Research in the area covering the efficacy of various types of meditation will prove to be useful in mapping subsequent developments in meditative design methods, and where they might sit on the DD. In turn, such practices can be integrated into design practice. Similar analysis may be applied to MP, or any other meditative design process. Therefore, the DD acts as scaffolding of the MDDF, and future development of meditative design methods can be conducted with specific phases in mind.

Discussion

Through considering the literature, examples of praxis, and the MDDF, the rich ground which is meditative design is apparent. Educators of meditative design methods should develop capacities to guide learners through a post-human-centered MDDF. As an educator, practitioner, and researcher, the praxis of meditative design is one where the embodied process of guiding meditation and developing mindfulness is necessary to not only conceive novel methods, but facilitate the journey of learners in their own praxis. Self-guided techniques can be learned through guidance. The indigenous Maori of the Whanganui river (a legal person) have the saying, "I am the river and the river is me". Such attitudes have been cultivated through a lived experience with the river coming from the local people's cosmovision (Collins & Esterling, 2019). This idea then connects with Akama's (2012) contention that mere utilitarian appropriations are not sufficient. The origins and holistic nature of meditative practices are apparent and the diverse indigenous knowledge should be respected and studied. Therefore, it is the responsibility of educators across disciplines to be mindful of their own perspectives and have the openness to adapt to the learners' needs in guiding meditative design processes. In terms of praxis, the framework needs flexibility to accommodate the development of learners and educators alike.

MDDF and the Between-Ness

The strategy of clustering methods into phases has useful precedent. The toolkits can indicate which phase a particular method may be useful in, such as those originally produced by the MediaLAB Amsterdam (2015). Simultaneously, as discussed earlier, each method may have multiple uses in different phases of the problem and solution space; one should avoid pigeonhole methods. Though covering the 4 phases of the DD may seem comprehensive, there is one issue with the current proposal. As an approach to research and develop meditative design, the use of the DD framework is project centric. In terms of a curriculum fit, a practice-based design program will be able to include such methods into workshop or project-based learning formats. However, when considering a more continuous, or co-design, co-evolutionary approach, this may lead to an understanding that meditative approaches are merely to be adopted when in "work mode". Further, what about meditative approaches *Ma* or between-ness. Such projects/briefs deserve further attention to consider inter-project relationships using MDDF.

How can we have empathy to design for complex ecosystems, when they are hyperobjects (Morton, 2013). MDDF methods such as MI and MP support such development of imaginative empathy. The type of post-human-centered thinking requires empathy for participants in our biosphere such as mallards, gut bacteria, or rivers. If we can harness radical empathy as an engine for creative design, it may drive motivation to develop novel and impactful solutions. According to Wright and Monsour (2020, p. 164) there is hope: "pedagogical practices such as reflection, meditation and contemplative practices and activities engage candidates with the world, with others, to foster awareness for living mindfully in ecological, and socially just ways".

Conclusion

A novel "Meditative Design Development" framework along with meditative design methods have been proposed in this chapter. These will be useful to educators across disciplines, who may wish to adopt, replicate, or adapt the described methods. The framework will assist in navigating how to find further opportunities to test meditative design techniques. Research opportunities at the intersection of meditation and design are identified as part of the literature review. The results of the cases discussed in this chapter add to the body of research in applied meditative practice in design, engineering, and innovation studies more broadly. From a co-design perspective, the methods can be tailored to use less jargon and be delivered in multiple languages, both online and offline. The praxis documented here should be taken as a work in progress and is an open invitation to collaboratively explore the rich meditative research space. It is hoped that the methods and the framework may allow educators and learners to approach the Anthropocene with radical empathy and boundless creativity, such as that which has been exhibited through the Whanganui River becoming a legal person.

Acknowledgements The research I have conducted for this chapter is partially supported by KAKEN grant no. 18K18315 "Toward a Convivial Design model with the Whanganui River, a legal person".

References

Akama, Y. (2012). A 'way of being' in design: Zen and the art of being a human-centred practitioner. *Design Philosophy Papers, 10*(1), 63–80.

Akama, Y. (2014, October). Attuning to Ma (between-ness) in designing. In *Proceedings of the 13th participatory design conference: short papers, industry cases, workshop descriptions, doctoral consortium papers, and keynote abstracts* (Vol. 2, pp. 21–24).

Akama, Y., & Light, A. (2015, April). Towards mindfulness: Between a detour and a portal. In *Proceedings of the 33rd Annual ACM Conference Extended Abstracts on Human Factors in Computing Systems* (pp. 625–637). ACM.

Andrahennadi, K. C. (2019). Mindfulness-based design practice (MBDP): A novel learning framework in support of designers within twenty-first-century higher education. *International Journal of Art & Design Education, 38*(4), 887–901.

Ball, J. (2019, October 1). *The double diamond: A universally accepted depiction of the design process.* https://www.designcouncil.org.uk/our-work/news-opinion/double-diamond-universally-accepted-depiction-design-process

Chimiklis, A. L., Dahl, V., Spears, A. P., Goss, K., Fogarty, K., & Chacko, A. (2018). Yoga, mindfulness, and meditation interventions for youth with ADHD: Systematic review and meta-analysis. *Journal of Child and Family Studies, 27*(10), 3155–3168.

Collins, T., & Esterling, S. (2019). Fluid personality: Indigenous rights and the Te Awa Tupua (Whanganui River Claims Settlement) Act 2017 in Aotearoa New Zealand. *Melbourne Journal of International Law, 20*(1), 197–220.

Colzato, L. S., Szapora, A., & Hommel, B. (2012). Meditate to create: The impact of focused-attention and open-monitoring training on convergent and divergent thinking. *Frontiers in Psychology*, 116.

de la Rosa, J., Ruecker, S., & Nohora, C. G. (2021). Systemic mapping and design research: Towards participatory democratic engagement. *She Ji, 7*(2), 282–298.

Draaijer, S. (2009). Improving the design process for selected-response questions in higher education. In *The Stellar Doctoral Consortium EC-TEL 2009.*

Escobar, A. (2018). *Designs for the pluriverse.* Duke University Press.

Goldberg, S. B., Tucker, R. P., Greene, P. A., Davidson, R. J., Wampold, B. E., Kearney, D. J., & Simpson, T. L. (2018). Mindfulness-based interventions for psychiatric disorders: A systematic review and meta-analysis. *Clinical Psychology Review, 59*, 52–60.

Guilford, J. P. (1956). The structure of intellect. *Psychological Bulletin, 53*(4), 267–293.

Henriksen, D., Richardson, C., & Shack, K. (2020). Mindfulness and creativity: Implications for thinking and learning. *Thinking Skills and Creativity, 37*, 100689.

Kabat-Zinn, J. (1982). An outpatient program in behavioral medicine for chronic pain patients based on the practice of mindfulness meditation: Theoretical considerations and preliminary results. *General Hospital Psychiatry, 4*(1), 33–47.

Kabat-Zinn, J. (2003). Mindfulness-based interventions in context: Past, present, and future. *Clinical Psychology: Science and Practice, 10*(2), 144–156.

Karalis, T., & Raikou, N. (2020). Teaching at the times of COVID-19: Inferences and implications for higher education pedagogy. *International Journal of Academic Research in Business and Social Sciences, 10*(5), 479–493.

Lebuda, I., Zabelina, D. L., & Karwowski, M. (2016). Mind full of ideas: A meta-analysis of the mindfulness–creativity link. *Personality and Individual Differences, 93*, 22–26.

Levy, D. M., Wobbrock, J. O., Kaszniak, A. W., & Ostergren, M. (2012). The effects of mindfulness meditation training on multitasking in a high-stress information environment. In *Proceedings of Graphics Interface 2012* (pp. 45–52).

Luberto, C. M., Shinday, N., Song, R., Philpotts, L. L., Park, E. R., Fricchione, G. L., & Yeh, G. Y. (2018). A systematic review and meta-analysis of the effects of meditation on empathy, compassion, and prosocial behaviors. *Mindfulness, 9*(3), 708–724.

MediaLAB Amsterdam. (2015). *Design methods—toolkit*. https://medialabamsterdam.com/blog/project/design-method-toolkit/

Morton, T. (2013). *Hyperobjects: Philosophy and ecology after the end of the world*. University of Minnesota Press.

Nakagawa Project. (2022, March 30). *Nakagawa project*. Strategic Design Course, Department of Design, Graduate School of Design, Kyushu University. https://www.sd.design.kyushu-u.ac.jp/en/project/nakagawa-project/

Niedderer, K., Tournier, I., Coleston-Shields, D. M., Craven, M., Gosling, J., Garde, J., Salter, B., Bosse, M., & Griffioen, I. (2020). Designing with and for people with dementia: developing a mindful interdisciplinary co-design methodology. In *IASDR 2017 Conference Proceedings*. University of Cincinnati.

Pahl, A.-K. (2005, August). A preliminary report investigating team creativity using meditation as a template for the co-evolutionary design process. In *Proceedings of the International Conference on Engineering Design*, Melbourne, August 15–18 (pp. 1–13).

Papanek, V. (1988). The future isn't what it used to be. *Design Issues, 5*(1), 4–17.

Rockström, J., Steffen, W., Noone, K., Persson, Å., Chapin, F. S., Lambin, E., Lenton, T. M., Scheffer, M., Folke, C., Schellnhuber, H. J., Nykvist, B., de Wit, C. A., Hughes, T., van der Leeuw, S., Rodhe, H., Sörlin, S., Snyder, P. K., Costanza, R., Svedin, U., ... Foley, J. (2009). Planetary boundaries:

Exploring the safe operating space for humanity. *Ecology and Society, 14*(2). http://www.jstor.org/stable/26268316

Rojas, F., English, S., Young, R., & Spencer, N. (2015, June). Making mindfulness explicit in design education. In *Proceedings of the 3rd International Conference for Design Education Researchers* (p. 623).

Rojas, F., English, S., Young, R., & Spencer, N. (2016, July). Bridging mindfulness and design. In *Proceedings of the 5th PhD in Design Research Meeting*. UD16-SUR/VIVAL.

Sanders, E. B. N., & Stappers, P. J. (2012). *Convivial toolbox: Generative research for the front end of design*. BIS Publishers.

Sanders, E. B. N., & Stappers, P. J. (2014). From designing to co-designing to collective dreaming: Three slices in time. *Interactions, 21*(6), 24–33.

Schwartzman, R. (2020). Performing pandemic pedagogy. *Communication Education, 69*(4), 502–517.

Sibinga, E., Webb, L., Ghazarian, S. R., & Ellen, J. M. (2016). School-based mindfulness instruction: An RCT. *Pediatrics, 137*(1).

Simpson-Little, D., & Long, C. (2010). Find the gap: Developing abilities in space consciousness for enhanced learning. In *DS 62: Proceedings of E&PDE 2010, the 12th International Conference on Engineering and Product Design Education—When Design Education and Design Research Meet*, Trondheim, Norway (2010).

Thomas, V., Remy, C., & Bates, O. (2017, June). The limits of HCD: Reimagining the anthropocentricity of ISO 9241–210. In *Proceedings of the 2017 Workshop on Computing Within Limits* (pp. 85–92).

Van Woerde, A. (2018). *Global goals jam edition*. https://www.marineterrein.nl/agenda/online-global-goals-jam-edition/

Visser, F. S., Stappers, P. J., Van der Lugt, R., & Sanders, E. B. (2005). Contextmapping: Experiences from practice. *CoDesign, 1*(2), 119–149.

Wright, M. F., & Monsour, F. A. (2020). Sustainability education assessments in teacher education: Addressing "wicked problems" in a postmodern world. *New Directions for Teaching and Learning, 161*, 155–176.

Zenner, C., Herrnleben-Kurz, S., & Walach, H. (2014). Mindfulness-based interventions in schools—A systematic review and meta-analysis. *Frontiers in Psychology, 5*, 603.

CHAPTER 16

Capacities for the New World: Changing Roles and Responsibilities of Educators and Learners

Isabelle M. M. J. Reymen, Jan D. Vermunt, and Chantal H. T. A. Brans

Global forces are impacting the higher education sector and profound change is imminent: the rise of continuous learning, the changing world of work, the blurring of industry boundaries, evolving digital behaviour, and increasing international competition (Ernst & Young, 2018; Graham,

I. M. M. J. Reymen (✉) · C. H. T. A. Brans
TU/e innovation Space, Eindhoven University of Technology, Eindhoven, The Netherlands
e-mail: i.m.m.j.reymen@tue.nl

C. H. T. A. Brans
e-mail: c.brans@tue.nl

J. D. Vermunt
Eindhoven School of Education, Eindhoven University of Technology, Eindhoven, The Netherlands
e-mail: j.d.h.m.vermunt@tue.nl

M. J. Lehtonen et al. (eds.), *Design Education Across Disciplines*,
https://doi.org/10.1007/978-3-031-23152-0_16

2018; Kumar et al., 2018). It is the moment for education to enter the twenty-first century: innovations in education need to be accelerated to address the grand challenges in our society more adequately. This holds particularly for education in which design is an important element, such as engineering education.

Graham indicates that the distinctive educational features of a new generation of engineering programs include "work-based learning, multidisciplinary programs and a dual emphasis on engineering design and student self-reflection" (Graham, 2018, p. iii). In her study, Graham (2018, p. 39) reports several trends, including "a move towards socially-relevant and outward-facing engineering curricula". Such curricula emphasize "student choice, multidisciplinary learning and societal impact, coupled with a breadth of student experience outside the classroom, outside traditional engineering disciplines and across the world" (Graham, 2018, p. iii). Another trend is "to deliver student-centred learning to large student cohorts through a blend of off-campus personalized online learning and on-campus hands-on experiential learning" (Graham, 2018, p. iv). Eindhoven University of Technology (TU/e), a leading university in science and technology educating future engineers, is in the process of making this transition towards more student-centred learning (Eindhoven University of Technology, 2018).

Since 2012, TU/e has made an impressive educational transition, with a forward-thinking perspective on engineering education through the implementation of the Bachelor College and Graduate School (Gommer et al., 2015). All TU/e graduates have a common engineering base and can develop a personal profile via a lot of elective space in their curriculum. They are stimulated to take more responsibility for their own learning, and the development of professional skills is more important (Gómez Puente, 2014). The Bachelor College is a genuine, curriculum-wide, systemic reform affecting every course, student, and educator (Graham, 2015).

To get to the next step in stimulating student learning and the creation of a societal relevant curriculum (Graham, 2018), we need students to develop new capacities. The learning of students in well-known inter-disciplinary student teams like Solar Team Eindhoven[1] serves as an

[1] https://solarteameindhoven.nl/.

example to TU/e and the innovative industry in the Brainport Eindhoven Region. The students in these teams demonstrate to learn, next to their curriculum, by working with passion on open challenges with a competitive edge. They learn to think at a systems level, collaborate with companies and society, experiment with users, learn to deal with failure, and show entrepreneurial behaviour. Because the students experience a learning urgency, their learning deepens. They develop and apply disciplinary knowledge and professional skills. Through TU/e innovation Space,[2] this innovative way of learning, called Challenge-Based Learning (CBL), is available to all students in their curriculum (Reymen, 2019).

Challenge-Based Learning is not a teaching method, but an educational concept; it is a view of how students are empowered to gain more ownership of their own learning. Working on challenges, like the UN Sustainable Development Goals, triggers students to acquire and apply new knowledge and skills. Students are offered challenges they are passionate about, and which make them learn based on intrinsic motivation. Since there is no answer known in advance to the open-ended challenges, educators learn together with the students, shifting attention from teaching to learning and from knowledge transfer to coaching. This way, students are guided by their coach and are in charge of their own learning path (self-directed and active learning). Learning to learn is a crucial competence to develop. CBL is often compared to design-based learning. A key difference is that in design-based learning, students often start from a problem and design a solution, whereas in CBL, a real world, open-ended societal challenge is the starting point of student learning (Membrillo-Hernández et al., 2019). TU/e innovation Space facilitates CBL for an increasing number of students across all programs and levels at TU/e, in courses and in extra-curricular activities. First pilots on the curriculum level are also started and therewith contribute to the educational vision of the university towards 2030 (Eindhoven University of Technology, 2018; TU/e innovation Space, 2020). Students really appreciate and request this type of change in education. CBL is a new step in transforming education in response to societal changes.

Collaboration with external stakeholders from industry, government, and society is a key aspect of working on challenges. Students learn in the ecosystems of the university, i.e. learn in the innovation hub. TU/e

[2] www.tueinnovationspace.nl.

is located at the heart of the Brainport Eindhoven Region. This is a high-tech region that functions as a growth accelerator for high-tech industry with a lot of interaction between research, industry, and society (Eindhoven University of Technology, 2018). This existing and strong collaboration in the region created a strong basis for TU/e to intensify the collaborations with industry and society towards education and education innovation. The work of Hoyos Bustamante (in this volume) nicely indicates how real-life projects with external stakeholders can be practically organized, via the role of a gregario, being a graduate student. The gregario has the essential role of a boundary spanner we also see needed in CBL. Via the gregario, the problems or challenges of industry and society are connected to the university. This leads to strengthen education and research at the university and the realization of alternative solutions for industry (Hoyos Bustamante in this volume). The collaboration between the university and companies can be full of (legal) hurdles, but a key tip given by Hoyos Bustamante (in this volume) in this section is, based on the approach of knowing through doing, "the most important thing is joint action".

When transforming education, reflection on the changes we make should not be forgotten. The study of Williams (in this volume) nicely indicates that to radically change higher education and our own practices, we require continuous attention to both action and reflection. And they indicate both tutors and students need to participate in change as learners. We love one of their final quotes: "It is vital that tutors and students learn with change, not just about it". This also means we need to create conditions for educators to tolerate the creation of uncertainty and unpredictability (Williams in this volume). The ability to deal with uncertainty (Reymen et al., 2015; Sarasvathy, 2001) and an entrepreneurial mindset (McGrath & MacMillan, 2000) is indeed key and requires more attention among educators and students.

The next section focuses explicitly on the role of design and designers in developing the capacities needed by students and educators to address the grand challenges in our society. Subsequently, we detail the changing role and responsibilities of learners and educators.

Capacities for the New World and the Role of Design and Designers

Which role can design and designers play in capacities for the new world? When going back to the definition of Herbert Simon (1996), design is the core competence for all professional activities. It is using knowledge to create what should be, things that do not yet exist; it is the activity of changing existing situations into desired ones. So, design is a key capacity for all professionals and thus also for all learners.

We believe all students need to develop capabilities to create change, and thus to be able to design. This means that design competencies are key competencies for all learners, in design, in engineering, but also even in broader education, if we want to prepare students as change agents of the future (see also Inamura's chapter in this book).

Educating learners with design capacities stimulates them to create things that were not yet there, come up with many new solutions for a problem that at the start is also not yet clear (like wicked problems). Instead of integrating other disciplines in design education (as suggested by Yeo in this volume), we propose to integrate design in the other disciplines, or at least in all study programs, to empower all students with design competencies.

Designers coming from specific design education can then be the experts in designing, and even advancing the design discipline. They can be included in teams when specific, more advanced, design expertise is necessary. We hereby invite the design professionals to make explicit their specific core competencies, in addition to the basic design competence necessary for all professionals, and their possible future roles. Williams (in this volume) has already made a very good contribution for the field of fashion design.

Changing Role and Responsibilities of Learners

The discussion so far implies rather drastic changes in the roles and responsibilities of learners in innovative pedagogical teaching–learning models in higher education. The type of learning that is being fostered in these models is authentic, intrinsically motivated, active, deep, autonomous, self-regulated, collaborative, interdisciplinary, meaning-oriented, application-oriented, and competence-centred in nature. This makes changes necessary in five components of student learning: learners' conceptions of learning, learning motivations and emotions, regulation of learning, depth of processing of content, and social aspects of learning (cf. Vermunt & Donche, 2017).

First of all, learners need to develop a conception of learning in which learning is seen as the active construction of their own personal knowledge instead of the intake of knowledge provided by others. They have to view it as their own responsibility to construct that knowledge and take full ownership of this process of personal knowledge construction. In this constructive conception of learning, it is useful that educators support learners to create this personal knowledge and good teaching equates to providing enough support for learners to enable their personal learning trajectory.

Second, learners need to be(come) intrinsically motivated to engage in active, deep, and self-regulated learning. The underlying principle of many innovative pedagogies is exactly focused on this. The opportunity to work on their own or chosen real-life problems, cases, or challenges aligns very well with their motivation as to why they have chosen a particular discipline or domain of studies, which gives a strong boost to invest energy in trying to solve the problems of work on the challenges. Moreover, research has shown again and again that intrinsic motivation is highly correlated with self-regulation and deep approaches to learning. Extrinsic motivation, with its focus on passing exams, has been shown to be highly correlated with reliance on teacher regulation of learning and surface approaches to learning.

Third, a very central feature of innovative pedagogies in higher education is active and self-regulated, self-directed learning. Learners need to make their own choices about what, how, and where to learn. Learners are in the lead of their own learning. They need to be able to identify their learning interests, recognize their learning needs, set learning objectives, plan a learning trajectory, monitor their learning progress, diagnose the cause of difficulties that arise on the way, adjust their learning accordingly, persist when needed, evaluate the success of their learning, and reflect on learning and their own role and responsibility in that endeavour. This is opposite to their role in traditional pedagogical models, in which the educator decides what, how, and when learners should learn and how the success of that learning should be assessed.

Fourth, innovative pedagogical models are more explicitly aimed at learning deeply instead of superficially. Learners are asked to engage critically with what they learn, to think for themselves, compare and contrast theories, cases, examples, and viewpoints, derive their own conclusions, take a stance in societal debates, relate what's new to them to what they already know, and aim to gain a deep understanding of their subjects and creative solutions for their challenges. This contrasts with traditional models in which assessment was often based on how well learners could remember what they had learned, for which a surface approach to learning would do.

Finally, collaborative learning is much more important in innovative pedagogical models than in traditional teaching in higher education. Challenge-based learning, project-based learning, and design-based learning all expect students to work and learn in small teams of typically 3–5 learners. They have to share the regulation of their learning, divide the tasks, balance working on the task with learning from the task, deal with free rider behaviour from team members, and balance leadership roles in the team. The learning outcomes are focused on the team process, the product, and the individual learning. For many learners this shift from competitive and individual learning to collaborative learning, in which they learn from and with each other, is a very new experience that they must learn. Therefore, scaffolding is key.

Changing Roles and Responsibilities of Educators

Not only the roles and responsibilities of learners change substantially in innovative teaching–learning models in higher education. Educators' roles and responsibilities show a parallel change. In traditional lecture-based teaching, educators' most important responsibilities are to motivate students, to organize and direct their learning, and to explain the subject matter clearly. However, to foster authentic, intrinsically motivated, active, deep, autonomous, self-regulated, collaborative, interdisciplinary, meaning-oriented, application-oriented, and competence-centred learning in their students, educators need to fulfil other and different teaching roles as well.

In assignment-based teaching for example, educators need to be able to make activating assignments, set students to work effectively, monitor their students' progress, and give feedback on students' work. In problem-based learning, an educator has to act as block coordinator, problem designer, skills trainer, and tutor of tutorial groups. Project-based learning requires educators to coach the collaboration between team members in a project group, supervise and guide project teams, and deal with free rider behaviour of individual students. In competency-centred teaching, educators need to fulfil roles like professional growth consultant, competency assessor, and study career advisor. In work-based learning educators need to be able to help students clarify their learning needs, design authentic tests, and to act as portfolio advisors and mentors (e.g. Vermunt, 2007).

In Challenge-Based Learning educators have many of the responsibilities discussed so far. On top of those, specific for CBL, they need to be able to choose and adapt educative challenges, coach students in how to learn from working on those challenges, help students to define doable subtasks, coach students in balancing working and learning goals, supervise their choice of just-in-time disciplinary study, collaborate with educators from other disciplines, provide expertise to students, and assess the quality of the challenge solution together with individual students' learning accomplishments.

In all these innovative teaching–learning models, educators get new roles like activator, model, challenger, diagnostician, evaluator, and reflector of students' individual and team learning processes. In addition, they may get new roles as team educator, developer of technology-enhanced active learning programs, etc. For many university educators these are new roles and they may find it difficult to adjust to the new responsibilities that these innovations ask from them. The transitions are not simply a matter of learning new teaching skills but require a transformation of their view of what good learning essentially is and what it means for education and educators to support such learning: a development from a main responsibility of organizing a study program and directing student activities to one that includes a high level of student autonomy, collaborative learning, and unpredictability.

Conclusion

To conclude, the transformation in higher education implies changing roles and responsibilities for learners and educators. In order to address the grand challenges in our society, learners need other capacities, of which design competence is a key one. Challenge-Based Learning is an educational concept developed as a step in transforming education in response to societal changes, where students are empowered to gain more ownership of their own learning and where the student journey is taken as a point of departure.

References

Eindhoven University of Technology. (2018). *TU/e strategy 2030; drivers of change*. Eindhoven University of Technology.

Ernst & Young. (2018). *Can the universities of today lead learning for tomorrow? The university of the future*. Ernst & Young Australia. https://assets.ey.com/content/dam/ey-sites/ey-com/en_au/topics/government-and-public-sector/ey-university-of-the-future-2030.pdf

Gómez Puente, S. M. (2014). *Design-based learning : Exploring an educational approach for engineering education*. Eindhoven University of Technology.

Gommer, E. M., Klaassen, R. G., & Brans, C. H. T. A. (2015). *Comparing Bachelor Curriculum Innovations at the Three Technical Universities—an exploratory study*. 3TU Centre for Engineering Education.

Graham, R. (2015). *Bachelor college impact study*. https://www.rhgraham.org/page-2/tu-e/

Graham, R. (2018). *The global state of the art in engineering education*. Massachusetts Institute of Technology (MIT).

Kumar, S., Gokhale, A., Bhattacharya, S., & Mathai, V. (2018). *University of the future: Bringing education 4.0 to life* (p. 60). FICCI, Ernst & Young LLP.

McGrath, R. G., & MacMillan, I. C. (2000). *The entrepreneurial mindset: Strategies for continuously creating opportunity in an age of uncertainty* (Vol. 284). Harvard Business Press.

Membrillo-Hernández, J., Ramírez-Cadena, M. J., Martínez-Acosta, M., Cruz-Gómez, E., Muñoz-Díaz, E., & Elizalde, H. (2019). Challenge based learning: the importance of world-leading companies as training partners. *International Journal on Interactive Design and Manufacturing*, *13*(3), 1103–1113.

Reymen, I. M. M. J. (2019). *Collaborate with social and economic impact*. Eindhoven University of Technology. https://research.tue.nl/nl/publications/collaborate-with-social-and-economic-impact

Reymen, I. M. M. J., Andries, P., Berends, H., Mauer, R., Stephan, U., & van Burg, E. (2015). Understanding dynamics of strategic decision making in venture creation: A process study of effectuation and causation. *Strategic Entrepreneurship Journal*, *9*(4), 351–379.

Sarasvathy, S. D. (2001). Causation and effectuation: Toward a theoretical shift from economic inevitability to entrepreneurial contingency. *Academy of Management Review*, *26*(2), 243.

Simon, H. A. (1996). *The sciences of the artificial* (3rd edn.). MIT Press.

TU/e innovation Space. (2020). *Vision TU/e innovation space*. TU/e Innovation Space.

Vermunt, J. D. (2007). The power of teaching-learning environments to influence student learning. *British Journal of Educational Psychology Monograph Series II, 4*, 73–90.

Vermunt, J. D., & Donche, V. (2017). A learning patterns perspective on student learning in higher education: State of the art and moving forward. *Educational Psychology Review, 29*(2), 269–299.

CHAPTER 17

Conclusion

Tomi Kauppinen and Laura Sivula

Design Interventions

As part of this collaborative journey as educators, we are (re)designing teaching and learning as outward facing and socially responsive. We need to create solutions at all levels, from the level of individual students and courses all the way to systemic changes in how we see education and how we lead higher education institutions. We also need many forms and depths of design pedagogies and transdisciplinary collaboration to build inclusive communities of learners equipped with not only knowledge but broader skill sets for the future. Rather than letting the role of universities diminish in solving wicked problems, we need to amplify the ways in which universities contribute to society.

T. Kauppinen (✉) · L. Sivula
Aalto University, Espoo, Finland
e-mail: tomi.kauppinen@aalto.fi

L. Sivula
Aalto University Executive Education Ltd., Helsinki, Finland
e-mail: laura.sivula@aalto.fi

M. J. Lehtonen et al. (eds.), *Design Education Across Disciplines*,
https://doi.org/10.1007/978-3-031-23152-0_17

> Complexity in global challenges demands a systemic, less siloed approach to create shared value from the stakeholder ecosystem. (Keane & Yeow, Chapter 4)

And yet, for collective change to happen, we need sensemaking through ambiguity, embracing the inherent future-creating nature of design in envisioning the future of teaching and learning. The conversations that run through this volume's chapters highlight a plurality of design interventions across disciplines and continents. We acknowledge design as one of the potential drivers behind transforming teaching and learning in higher education institutions. Due to design's future orientation and reliance on creative problem-solving skills, research on design thinking has been gaining momentum during the last decade or so. While there is a considerable body of research arguing for design thinking's positive impact in the classroom, this edited volume argues that design's impact goes beyond the classroom and the here-now.

Creativity, communication, collaboration, critical thinking, and spatial thinking are needed to both create sustainable futures and offer meaningful educational services and products in these futures. Learning, as well as an attitude and the capabilities for lifelong learning, are pivotal in achieving desirable futures. More specifically, and in line with arguments for design's role as a social and environmental catalyst, this volume brings together contemporary research on design education in and outside the design domain by illustrating how higher education institutions can be transformed to better respond to contemporary challenges related to environmental issues, social inequality, responsible business, and policy-making.

> There is an appetite to draw on a range of knowledges and to take a multi-faceted approach to pedagogy. (Williams, Chapter 12)

We identify five design interventions highlighted in this volume used to advance teaching and learning:

- Designing education that blurs disciplinary boundaries and eventually goes across disciplines
- Utilising design pedagogies to transform education to better address societal and environmental issues, ambiguity, and notions of inclusion and exclusion

- Collaborating with partners (e.g. public and private sector) in creating research and teaching initiatives to respond to contemporary and future challenges, building stronger connections between universities and their surrounding communities
- Recognizing the impact of meaningful learning experiences, including design pedagogies, human-centred approaches, challenge-based learning, and work-integrated learning
- Actionable methods, and use of tactile, tangible, and physical learning approaches as part of both online learning and studio classes.

Next, we will provide concluding remarks about how these design interventions, in practice, show up as drivers for change.

Drivers for Change

The world is getting more and more complex in the sense of its very definition. Higher education and the jobs we prepare our students for are no exception.

> In the multi-layered approach to organisational transformations, it seems that the approach in design pedagogy has much to offer when extrapolated to other parts of higher education and ideally will spark conversations that enable smoother transitions and innovations within universities that we can hopefully look forward to. (Chew, Chapter 6)

Let us think about what makes the world so complex and what the drivers are for accelerated change. The different complicated parts of the world are increasingly interrelated and connected in a multitude of ways. Information flows via social media and news outlets at an increasing speed. Services are running online and accessible from almost any part of the world. However, it is a world that is obviously not fair, or equal, for all. Online access is not guaranteed globally due to lack of infrastructure or because of firewalls set for often truly complex economic and political reasons. At the same time, we can evidence that most digital services are designed for some people, thus excluding others. Where are the digital services for the Global South?

> A new framework is needed that will allow all designers to work across sectors to develop the technologies they need to support business-related activities, solve the eternal problems of poverty, unemployment and inequality, and allow them to create a new world order in which humanity is placed above profit and power. (Yeo, Chapter 14)

Flight, train, ship, and other traffic routes carry people to almost any part of our global world. Global supply chains and logistics feed people and provide them with products designed in one place and manufactured in another. Data and visual analytics provide insight for designers and marketers to match their needs. Data is used to recommend people products, services, or media to consume. Design education can certainly react to change but it has the role of being a proactive change-maker for business, life, and society.

Visualisations and narratives guide our actions. We need to emphasise the role of ethics and philosophy as a foundation for design education.

> Today, designers need to value philosophy, with its emphasis on ethics, and integrate ethical ideas into a more purposeful design education. When critiqued and examined, philosophy can help learners navigate over-information and debunk misinformation. (Korenblat, Chapter 9)

All of these complexities challenge us: can we together see the world similarly and agree on core principles and observations?

> In the social constructivist theory, shared meaning among individuals creates the knowledge of reality we have. Our identity is created by the interaction we have with the environment and society and the reaction of the expectation of society. (Valencia et al., Chapter 7)

People still need to design, create and offer media, services, and products, although obviously increasingly with the help of computers and AI. Designers with their "can do" attitude and skills enable the running of iterative design processes.

> The advantage of design is the real time, actual experimentation. Taken together, both design and business modes of working may gain considerable experience from collaboration. (Kristensen & Gabrielsen, Chapter 10)

Higher education institutes need to employ design to reshape their offerings across disciplines and ensure learners can state with pride that they can take on diverse challenges and early on include users into design processes. A balance needs to be found between practical and theoretical skills. New ways of working call for the continuous learning of new skills. Social/soft skills like communication, collaboration, critical thinking, and creativity are the four Cs of the future of learning. People study or work shorter or longer periods of time out of their home country. Digital communication allows for keeping a connection to families and friends in other places on Earth, or in some rare cases also in space. As a result, it has brought dramatic changes to the lives of people and serves as a call to designers to adjust their work accordingly.

> ...socially responsive institutions can empower emerging designers for social change and the important transdisciplinary work ahead. (Brophy et al., Chapter 3)

The change has obviously already been going on for decades or even centuries and is now rapidly accelerating. Traditional jobs are changing fast, with many of them vanishing completely. Just think about how the printing industry has changed over the centuries, from being a highly skilled and manual job in the past to the fully automated processes for printing magazines and newspapers and to even on-demand book printing that we see today. Robotics, automation, and artificial intelligence together with global competition and awareness of products and services all provide thresholds but also opportunities for new and old players. Universities have a role in both educating and being forerunners both for and with society.

> Instead, we need to *reframe* existing conceptions; to see the same conception viewed from a different perspective or viewpoint. (Jones, Chapter 11)

Getting exposed to multitudes of viewpoints, approaches, opinions, and arguments while building the designer's identity and skills is a result of thoughtfully curated design education. One single field cannot clearly solve all challenges nor take opportunities to serve citizens while ensuring we will have an inhabitable planet, given the ongoing climate change. The future is also very uncertain and it is not so possible to predict what exactly will happen in the future. For this, it is vital to have a designer's attitude, to be sensitive to weak signals, and to be able to prepare well before change happens. Proactivity is essential in preparing for and shaping the future. However, this will not be easy, given how uncertain, ambiguous, and unpredictable the world can truly be.

> The pandemic therefore brought not only a focus on the utility of uncertainty and ambiguity in the design process and for design education; it highlighted a need for *all* educators to anticipate (and even embrace) ambiguity as they interpreted their teaching practices differently to continue supporting student learning. (Kelly, Chapter 5)

Indeed, we can and should learn a lot from the Covid-19 pandemic and how reactively higher education faced the new changed situation in which no students, nor faculty, could be on campus. The culture of being and acting as a designer is key but needs to happen across disciplines to make and prepare for changes. Transdisciplinary, grand challenges will need both transdisciplinary skills and teams to provide solutions.

The future can bring something that we cannot foresee, we can only guess and ask questions. Will space travel be an everyday business in a few decades time? How many robots will there be and what will be their roles in our everyday life? It is important here to have a perspective for learning that reaches well beyond this year or the next few years.

> It is of importance and urgency then, to empower learners in such a way that they can imagine and become actors to make the diverse impacts required. (Inamura, Chapter 15)

Will people be granted more free time due to the rise of automation, or will it be available only to some people? What will the class of people with more free time want to do with it; will they consume more and more digital services or will they want to reconnect with nature and other human beings? Is ever-increasing growth and consumption what people and nature need?

Pedagogies Changing Societies

Designing education and design education both call for novel pedagogical approaches. Human-centred approaches, challenge-based design scenarios, a wide range of communication methods, future-orientation, and transdisciplinary attitude covering a wide range of fields (like arts, humanities, psychology, philosophy, science, computer science, engineering, or business) are some of the key elements of pedagogies that prepare learners to be designers of the future. The key is to mix learners in teams from an early stage so that they learn together.

At the same time, designers need to create their own identity and offerings by learning core skills, forcing them to be the designers in those cross-disciplinary teams. Changing society for the better does not happen by accident. Design plays a vital role in orchestrating learning into meaningful and fruitful experiences.

> Key pedagogical practices that are found in Design-Based Learning include: learning situations framed through a critical stance; employing problem-focused design scenarios; using a range of alternative communication methods in the studio; embedding interdisciplinary and interprofessional learning opportunities often within collaborative and curricular and co-curricular experiential learning situations; and framing design as a hands-on, future-oriented and human-centred activity. (Rowe, Chapter 2)

Nature, human civilisation, society, and businesses now all face grand challenges. Incorporating challenges as starting points for learning will support ensuring that designers learn how to apply theoretical knowledge in practice. Instead of providing the students with often trivial assignments confined in the classroom, we need to blend the boundary between the classroom and the surrounding environment.

> Challenge-Based Learning is not a teaching method, but an educational concept; it is a view of how students are empowered to gain more ownership of their own learning. Working on challenges, like the UN Sustainable Development Goals, triggers students to acquire and apply new knowledge and skills. (Reymen et al., Chapter 16)

In practice this can mean that educators prepare project briefs based on digitally available descriptions of grand challenges and related resources and inspire students to feel ownership in aiming to solve them to the greatest extent possible within the learning context. We need to employ pedagogies that can prepare designers with the right skills and attitudes for the uncertain, ambiguous world they will face in their work-life.

We live in a society, among other people, machines, animals, nature, and our whole environment. Identifying challenges that call for solutions, exchanging ideas, and discussions, all happen via interaction where creating shared meanings and common ground is imperative. How can pedagogies enable us to create these "shared meanings" that will benefit society? This can happen by employing openness, trust, collaborative mindsets, agreements, frameworks, and formal requirements as the core elements.

> We apply what we call an open method: it is in continuous construction, emerging from the synergy between educators and students, with enterprises and communities. Consequently, in order to launch an interdisciplinary course, we begin with a voluntary agreement between the academy and the enterprise. (Hoyos Bustamante, Chapter 13)

Openness, trust, and collaboration will in practice mean the need to include society and its members in different roles (like citizens, customers, users) early on in design processes. Future-proof pedagogies need to prepare designers accordingly. When community-building is included as a core element in pedagogies, designers learn early on to count on others in all of their actions, design decisions, and life in general and provide similar support for others in the community. Indeed, having a context and community to work in provides interactions, feedback, and support. Learners will benefit from pedagogies ensuring that there is a meaningful and professional real-world context for their learning.

For instance, internships and other forms of collaborations with industry will prepare learners with skills, mindset, agility, and practical expertise. However, higher education has a key role in preparing learners to have a strong basis for moving to real-world professional environments. Learning needs to be designed to be a beneficial learning experience. Values are in the core of pedagogies that can truly change the world. Further, learners need to have active roles in their learning experiences so that their identity as professionals strengthens throughout their studies. Faculty will be there to assess learning but, very importantly, also to support.

To fully support learners, educators need to step down from their podiums to be with students and to create an environment that allows both students and faculty to give constructive feedback. Empathy and, more generally, emotions play a crucial role in being an approachable and supportive educator among learners.

Interacting with work-in-progress versions of designs, allowing students to improve them over time based on interactions and new learnings resembles a working-life experience where assessment is a continuous daily practice of ensuring high-quality services and products. Pedagogies need to ensure there are tactile, tangible, hands-on, multi-sensory elements as part of the learning experiences. Hands-on creation, in an iterative manner, will ensure there is always something to show and ask feedback for, as part of the learning journey. Online learning can and should include tactile and storytelling elements. After all, even online learning always happens in some place and time: the learner is situated in physical spaces (like home, studio, or campus spaces).

Finally, design pedagogies need to prepare students for a lifelong learning journey where skills and knowledge are continuously updated according to what is needed by design situations.

> As the embedding of WIL elements are discussed, there is the opportunity to rethink how pedagogy can benefit from explicitly creating curriculum and instruction seamlessly with work-integrated learning assets - but, many can say this already occurs in some form with the way the likes of Google and Amazon plan to disrupt college degrees by providing job-ready on demand learning opportunities. (Dickson-Deane et al., Chapter 8)

Considerations

In this edited volume, design education is framed as an inclusive, human-centred, and future-oriented activity transforming how we teach and learn. Looking at the depth and breadth of the examples of design as a discipline and design pedagogies in this volume sends out a clear message of the transformative power of approaches. As educators, we do need to re-evaluate and reinvent how we teach and interact with both students and the surrounding world. The examples of design interventions portrayed in this volume not only meet the immediate micro needs on an individual learner level but also address organisational needs on the meso level and societal needs on the macro level.

The modern higher education institution is forward-thinking and capable of making agile adaptations as the world changes. Developing leadership and managerial competencies in higher education institutions is often overlooked. Breaking through organisational inertia is a challenge for the modern educator. Disruptive practices are not always encouraged or do not have enough space in relation to burdening administrative duties. While this is not the case in all higher education institutions, there is a greater need for experimentation and the creation of novel educational opportunities that benefit both the student and the institution.

The challenge with current terminology is that it reinforces the hierarchical status between instructor and their students. However, as we consider learning to be an open-ended endeavour, the instructor also learns before, during, and after the process. While this claim is by no means novel, we think it is worthwhile to highlight both the student and the facilitator aspects side by side to fully comprehend the changes currently taking place in the higher education sector. For example, during the last few years, there has been a rising tendency to incorporate design-driven methodologies into disciplines outside the design school context, but so far the main arguments have been built upon design thinking, in itself a step-wise model. As mentioned in the beginning of this volume, we need to be careful to avoid the *transform washing* trap, making changes on the cosmetic level rather than advocating for change on all levels, ranging from the syllabus to the leadership.

To clarify, we are not advocating here that universities become vessels for immediate societal gains—on the contrary: with the design interventions presented and discussed before, our aim is to show and articulate a vision for higher education institutions. A vision that situated universities in relation to other societal actors: the public sector, corporations, individuals, and non-governmental organisations. And not only societal actors but also societal factors: environmental and social. This, we argue, is achieved when we start to approach courses and curricula from a co-learning perspective together with society and its stakeholders.

Discussion taking place in this edited volume focus on how and why universities can be integrated into the surrounding society by positioning themselves as sites for authentic experiences that are both interconnected and open-ended. What is more, as the contributions in this volume suggest, we see design interventions as an activity that provides the participating actors with opportunities to contribute to society.

Designing transformative learning experiences and orchestrating content production creates fascinating research questions to be included in the research agenda for the years to come. What are the educational elements that truly prepare learners to show their disciplinary skills and use creativity, collaboration, communication, and critical thinking to ensure societal usability and sustainability of their designs? This calls also for a more profound vision to integrate strategic pedagogical interventions into the operating models of higher education institutions for long-term effect.

Moreover, as we have highlighted throughout this volume, transformation of education is possible and can be done within and across disciplines. Transformation will need time and there are certainly obstacles (like lack of flexible physical and virtual spaces or old-fashioned one-way teaching cultures). However, the key is to focus on designing and creating authentic educational experiences with real-life relevance. Combining tactile activities with virtual ones, and theory with practice as a creative mix can support engaging learning experiences. Further on, educators have a role in actively shaping the future of higher education by initiating and maintaining close connections with surrounding communities, moving from reactive responses to proactively shaping the future of higher education.

We thank you for spending time with us via this volume. We wish you engaging and productive experiences in your journey with designing across disciplines.

Correction to: Strategy for Building a Transformative Relationship Between the Academy and the Social and Business Environment Through Interdisciplinary Work

Gloria Hoyos Bustamante

Correction to:
Chapter 13 in: M. J. Lehtonen et al. (eds.), *Design Education Across Disciplines*, https://doi.org/10.1007/978-3-031-23152-0_13

The original version of Chapter 13 was inadvertently published with incorrect chapter author's surname "G. H. Bustamante" which has now been corrected to "G. Hoyos Bustamante". The chapter has been updated with the change.

The updated original version of this chapter can be found at https://doi.org/10.1007/978-3-031-23152-0_13

M. J. Lehtonen et al. (eds.), *Design Education Across Disciplines*,
https://doi.org/10.1007/978-3-031-23152-0_18

Photo Credits

Figures 3.3 to 3.6—Claire Brophy.
Figure 6.1—JiaYing Chew.
Figures 7.1 to 7.7—Aldo Valencia.
Figures 8.1 and 8.2—Camilla Dickson-Deane.
Figure 9.1—Andrea Frank, Sarah Wyman and Joshua Korenblat.
Figures 9.2 and 9.3—Andrea Frank.
Figure 9.2 (bottom right)—David notMD.
Figures 15.1 to 15.3—Tokushu Inamura.

M. J. Lehtonen et al. (eds.), *Design Education Across Disciplines*, https://doi.org/10.1007/978-3-031-23152-0

Index

M. J. Lehtonen et al. (eds.), *Design Education Across Disciplines*, https://doi.org/10.1007/978-3-031-23152-0

The manufacturer's authorised representative in the EU is Springer Nature Customer Service Centre GmbH, Europaplatz 3, 69115 Heidelberg, Germany. If you have any concerns regarding our products, please contact ProductSafety@springernature.com

Printed and bound by CPI Group (UK) Ltd, Croydon, CR0 4YY
28/09/2026
02224072-0011